BASICS OF CHEST FILM INTERPRETATION

BASICS OF CHEST FILM INTERPRETATION

Cynthia B. Umali, M.D.
Assistant Professor of Radiology, University of Massachusetts Medical School;
Radiologist, University of Massachusetts Medical Center, Worcester, Massachusetts

LITTLE, BROWN AND COMPANY Boston/Toronto

First Edition

Library of Congress Catalog Card No. 87-82701

ISBN 0-316-88739-0

Printed in the United States of America

HAL

To my mother, who was my first and best teacher.

To Roel, Ryan, and Eloisa, for the constant love and support they have given me and for allowing me the luxury of free time to write this book.

To the residents, who through the years have taught me more than I can ever teach them. This book is the product of those countless, enjoyable hours we spent learning together and from each other.

To my friends for keeping faith in me.

PREFACE

In recent years, no field of medicine has exploded with as much progress as the field of diagnostic imaging. The development of new modalities (ultrasound, computerized axial tomography, and magnetic resonance imaging), however, has not obviated or diminished the need for plain film radiography. Plain films have remained as the first radiographic examination that physicians order, in most instances. A physician can use the radiograph as a diagnostic tool only after he or she has learned to recognize the normal and the abnormal and also what the abnormality represents. This is what we have done in this book on the basics of chest film interpretation: provide a step-by-step guide to film analysis and a working differential diagnosis for the abnormalities.

I would like to thank Jean Edmunds, Liz Shultis, Betsy Bohanan, and Sandra Kidd for their help in the preparation of this manuscript. My thanks also to Drs. Jim Coumas, Dick Waite, and Bob Carbonneau for their constructive comments after reviewing the manuscript and to Dr. Karen Reuter for her invaluable help during the early stages of this project. My very special thanks to Mary Cunion for the illustrations and Sandy Costa for the research materials.

C. B. U.

CONTENTS

BASICS OF CHEST FILM INTERPRETATION

STEP 1 EVALUATE LUNG AERATION

When looking at a chest film, most of us immediately focus on the lung parenchyma. Appropriately, standard chest film technique is geared to best evaluate the lung fields. Thus, this is the first area we will discuss.

The density of the hemithorax is influenced by the following five factors:

1. Lung volume, as determined by the number and size of alveoli.
2. Aeration of alveoli.
3. Blood flow.
4. Pleural integrity.
5. Chest wall composition.

The normal chest (Fig. 1-1), when viewed from apex to base, shows a progressive darkening, or increase in lucency, of the lung. This progressive darkening is best appreciated in the lateral view. The outlines, or borders, of intrathoracic soft tissues such as the heart, vessels, and diaphragm are well defined because of the interface between air density (lung) and water density (soft tissues). By answering several questions one by one, the viewer can make an objective gross assessment of the lungs.

1. Are the lungs of equal size and lucency? If not, the discrepancy can be explained by answering the following questions:
   - Is the discrepancy the result of the patient being rotated during positioning (Fig. 1-2)?
   - Is one lung smaller because of hypoplasia or volume loss from atelectasis or surgical resection? Or is one lung actually bigger from hyperaeration, as in a check-valve obstruction of a bronchus or emphysema?
   - Is the discrepancy caused by a relative difference in blood flow?
   - Is the discrepancy due to an overlying density such as pleural thickening, pleural fluid, or soft-tissue mass in the thoracic wall? The possible absence of a normal overlying structure, as seen following a mastectomy (Fig. 1-3), must be considered.

Fig. 1-1. *A*. Posteroanterior (PA) view of normal chest. Upper arrows trace the position of a normal minor fissure. Fissures are normally seen as very thin lines. Lower arrows trace the inferior accessory fissure. Well-defined cardiac borders are seen: right atrium (1), ascending aorta (2), superior vena cava (3), aortic knob (4), main pulmonary artery (5), left atrial appendage (6), left ventricle (7). A lower right hilum (8) as compared to the left (9) is seen. Hemidiaphragms (right, 10; left, 11) are well outlined. Costophrenic sulci (arrowheads) are sharp. *B*. Arrows point to the cardiac incisura; right ventricle (1), pulmonary outflow tract (2), ascending aorta (3), aortic arch (4), retrosternal air space (5), left atrium (6), left ventricle (7), right pulmonary artery (8), left pulmonary artery (9), right hemidiaphragm (10), left hemidiaphragm (11).

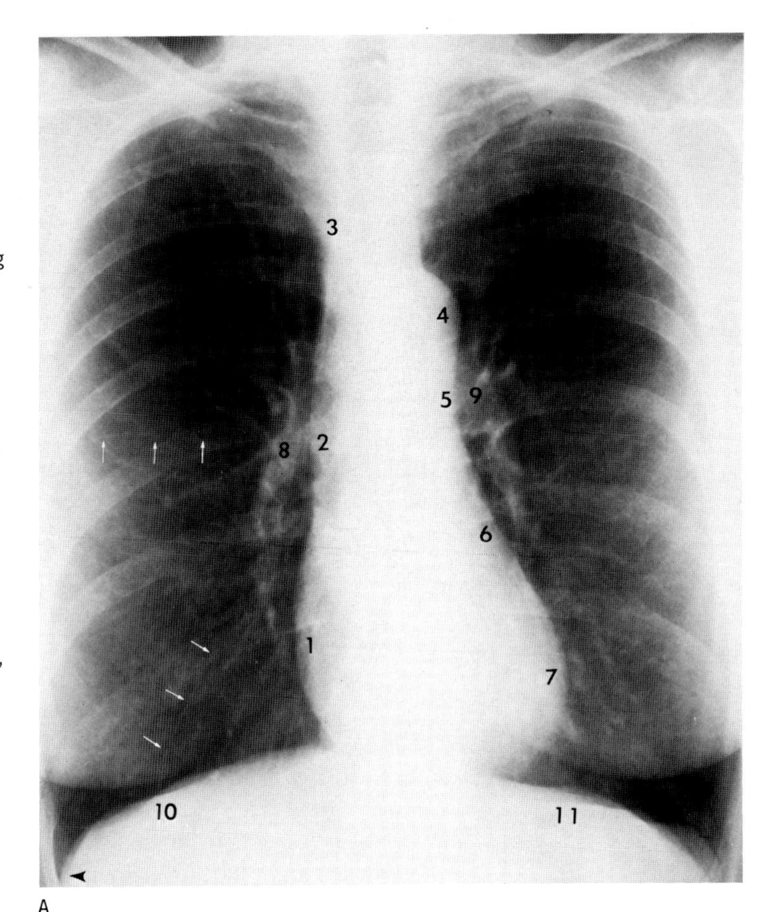

A

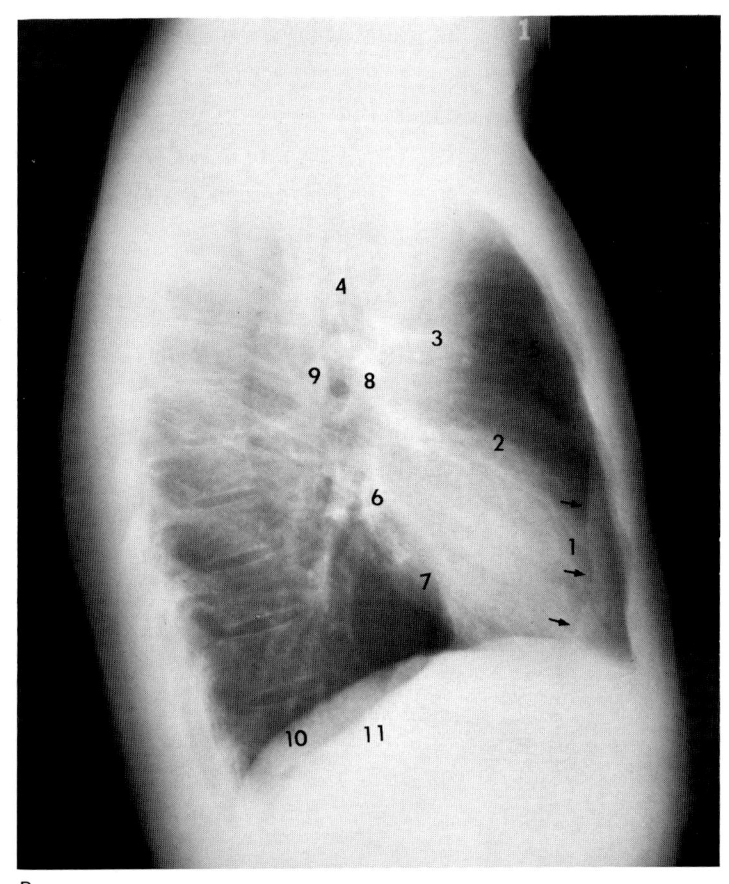

B

Fig. 1-2. Rotation producing a more lucent left hemithorax as compared to the right side.

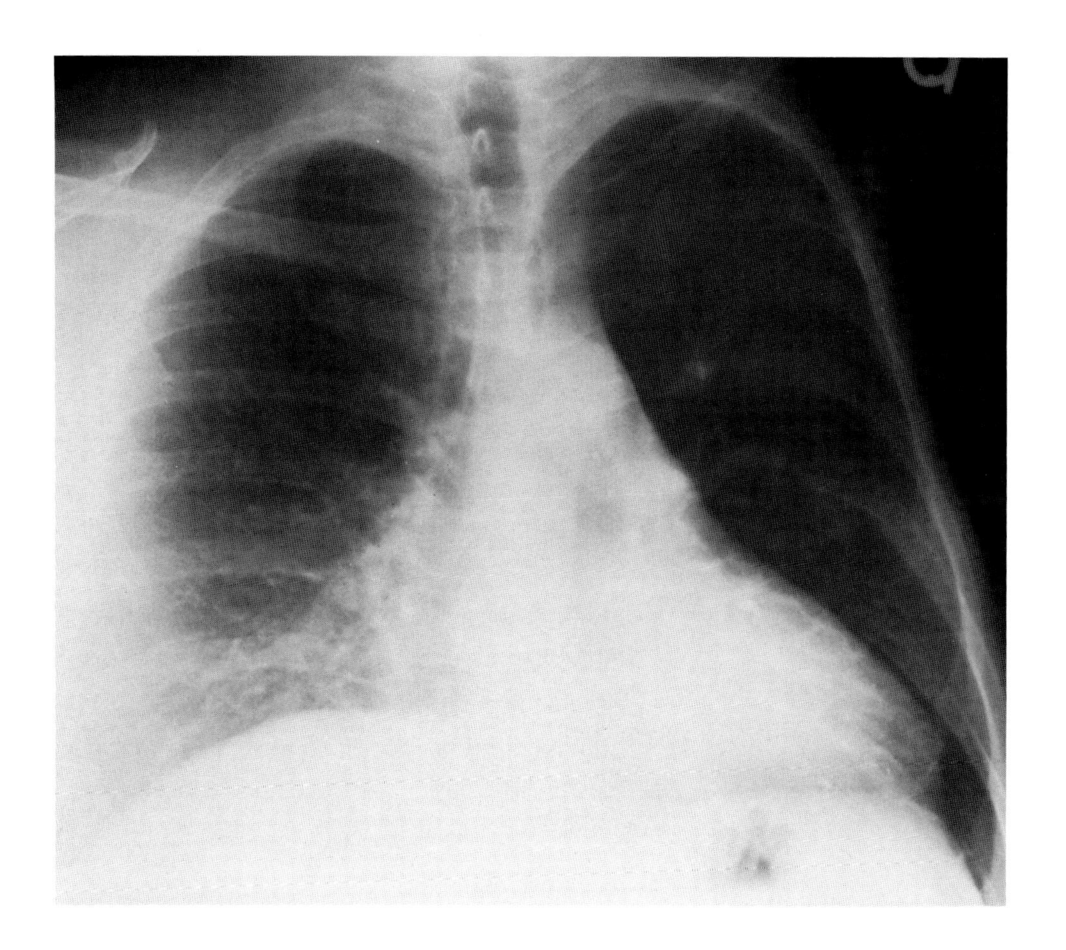

Fig. 1-3. Right lung (dark arrow) is more lucent than the left (open arrow). The patient had a right-sided mastectomy.

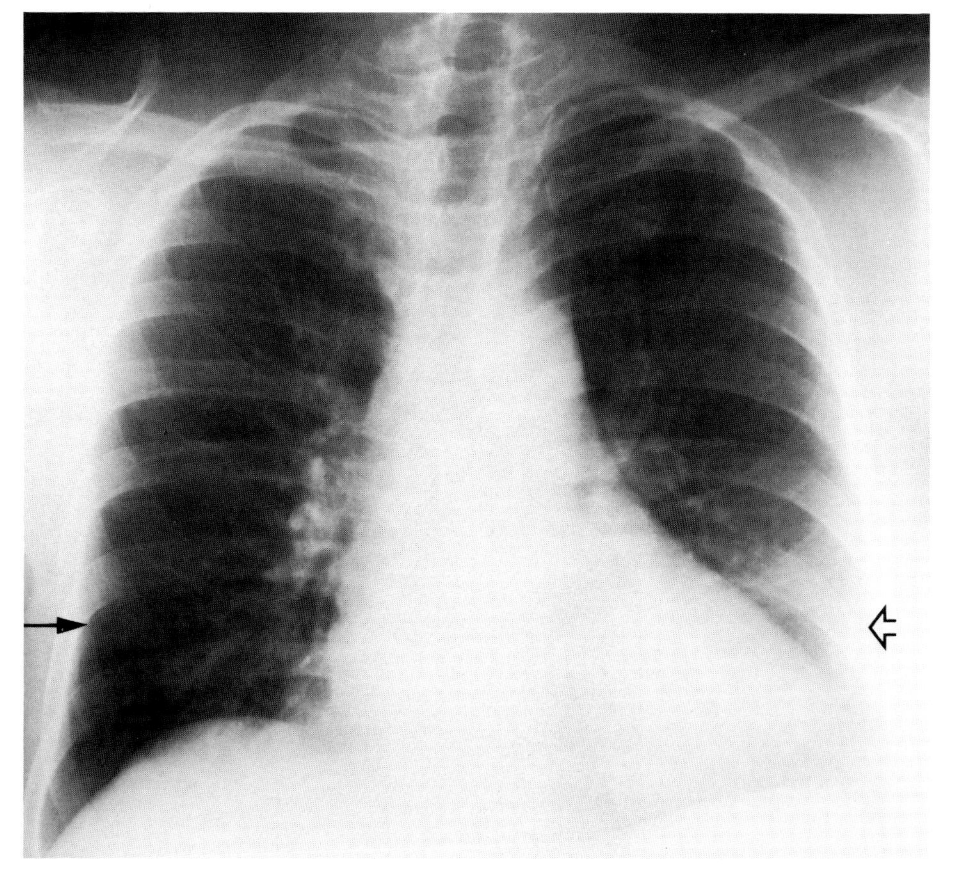

Is the discrepancy caused by the presence of air in the thorax outside the lung (pneumothorax) (Fig. 1-4)?

2. Do the lungs show a gradual but progressive increase in lucency when viewed from apex to base (except in areas overlapped by the breasts in females and the pectoralis muscles in muscular individuals)? If not, this should alert the viewer to the possible presence of abnormalities. Viewing this progression is most useful when evaluating a lateral film of the chest (Fig. 1-5).
3. Are the cardiac and diaphragmatic shadows homogeneous? Are the intervertebral disc spaces lucent (except where they are crossed by vessels or ribs)? If not, one should consider a possible abnormality (Fig. 1-6).
4. Are the cardiac borders, great vessels, and hemidiaphragmatic outlines well defined? If not, one should again consider abnormalities (see discussion of the silhouette sign).
5. Are the fissures, mediastinum, hila, and diaphragm normal in size and position (see discussion of individual areas)?

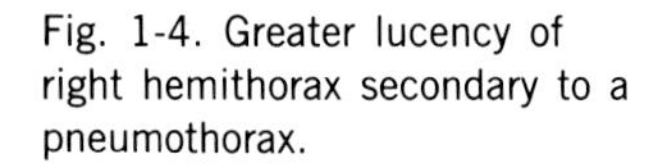
Fig. 1-4. Greater lucency of right hemithorax secondary to a pneumothorax.

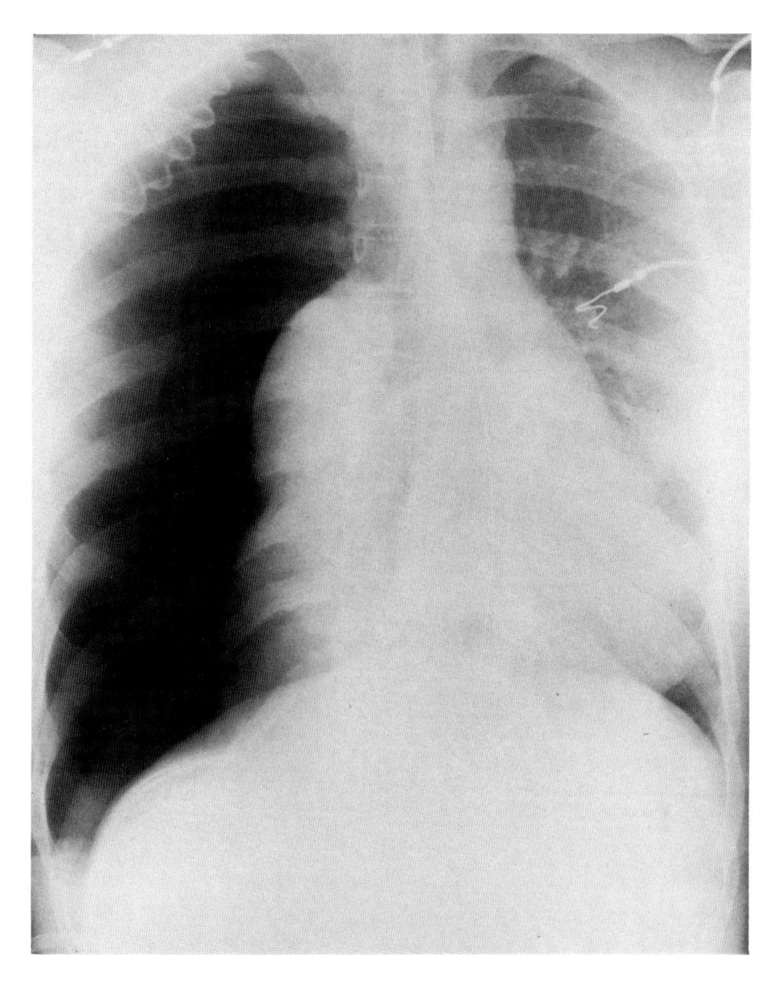

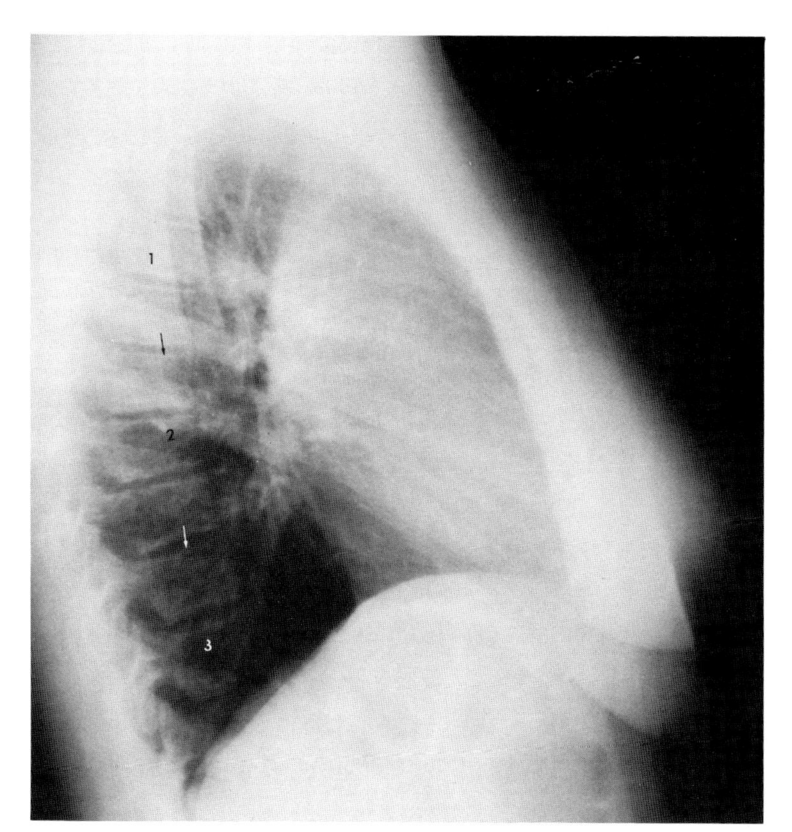

Fig. 1-5. A progressive increase in the blackness of the lung as one scans it posteriorly from top (1) to midportion (2) to bottom (3).

Fig. 1-6. *A.* Inhomogeneous density of cardiac silhouette with the lateral aspect (1) more lucent than the medial aspect (2). The increased density was produced by a posterior segment left-lower-lobe consolidation. Arrow points to edge of consolidation. *B.* Inhomogeneous cardiac shadow (1) same as in *A* but on the right side. Edge of the increased density, this time produced by atelectasis (2), is outlined by arrowheads.

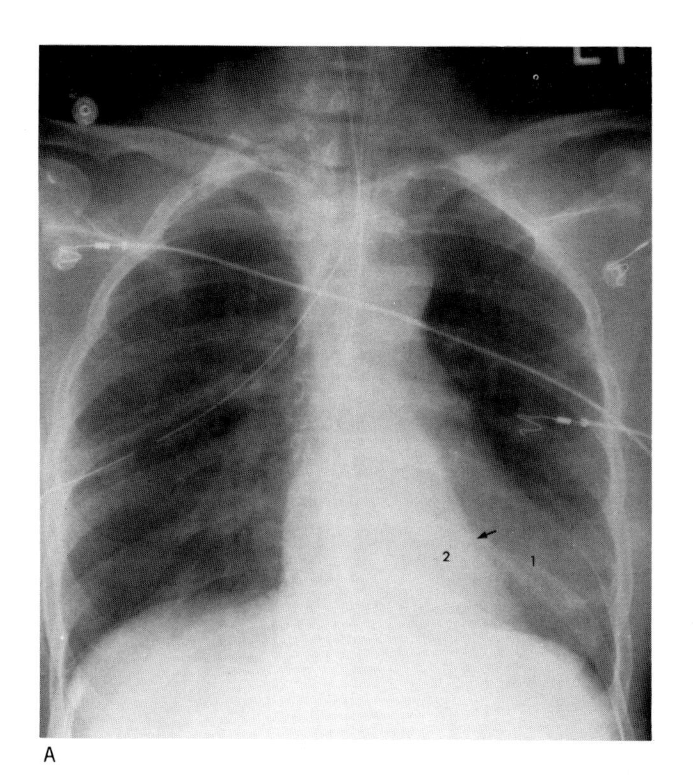

A

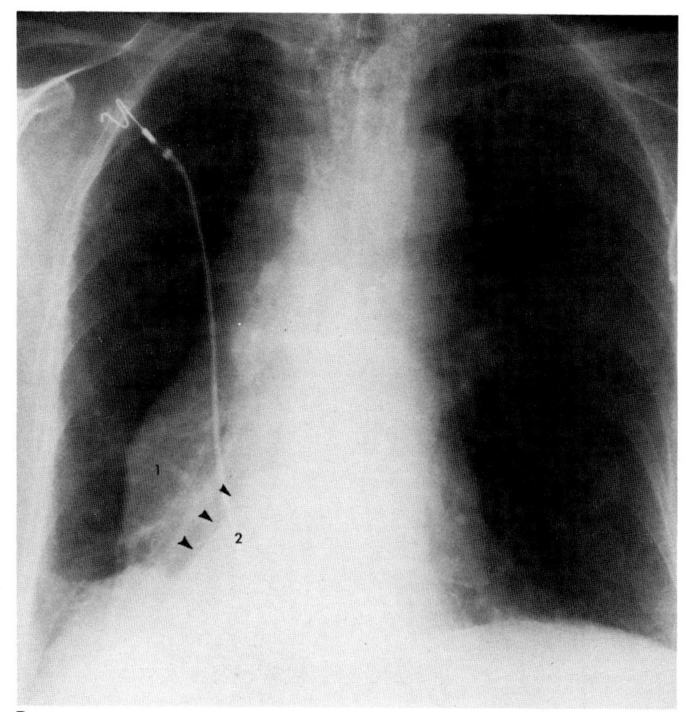

B

STEP 2

EVALUATE THE PARENCHYMA

Parenchymal densities have myriad causes. The physician must determine whether they represent consolidation, atelectasis, infarction, benign or malignant nodules, cavities or cysts, or interstitial disease. Then, too, an attempt should be made to localize the lesions.

Certain basic principles are followed in localizing intrathoracic densities. One is the location of the densities in relation to the fissures (Fig. 2-1). Above the minor fissure is the right upper lobe; between it and the major fissure is the right middle lobe; and below the major fissure is the lower lobe. On the left above the major fissure is the left upper lobe, and below, the lower lobe.

Another basic principle is the use of the *silhouette sign* [1]. This sign is based on the premise that we can recognize individual structures because of the differences in their densities. If a substance or a structure comes into anatomic contact with a structure of equal radiodensity, its previously visible individual borders are obliterated (Fig. 2-2). Thus, a water density (atelectasis, pneumonia, or mass) in the anterior aspect of the right upper lobe or anterior mediastinum may obliterate the ascending aorta border or part of the right cardiac border (Fig. 2-3). In the posterior portion of the left upper lobe or posterior mediastinum, portions of the descending aorta with or without the aortic knob may be obscured. In the right middle lobe or lingula, a portion of the heart border on the right and left sides, respectively, may be obliterated. And in the basal segments of the lower lobes, portions of the diaphragm may not be visible (Fig. 2-4).

An extension of the use of the silhouette sign is the *cervicothoracic sign* [2]. If a mass obliterates the borders of soft tissues of the neck, it must be cervicothoracic in location. Since the cephalic border of the anterior mediastinum ends at the level of the clavicles, if the defined border of a density stops at that level, it is anteriorly located; if it extends beyond the clavicles, it is posteriorly located (Fig. 2-5). The *thoracoabdominal sign* [2] is another extension of the silhouette sign. If the entire margin of a density is seen

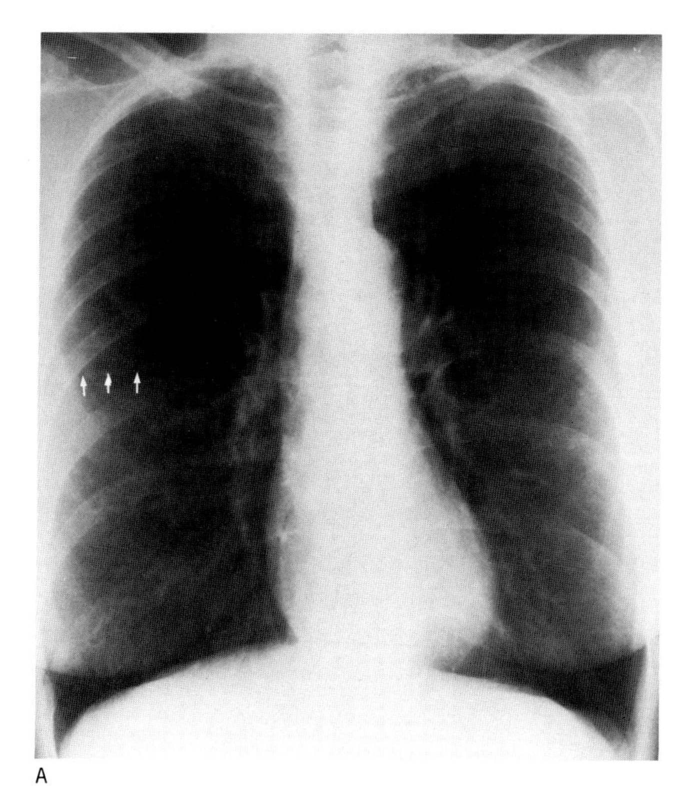

A

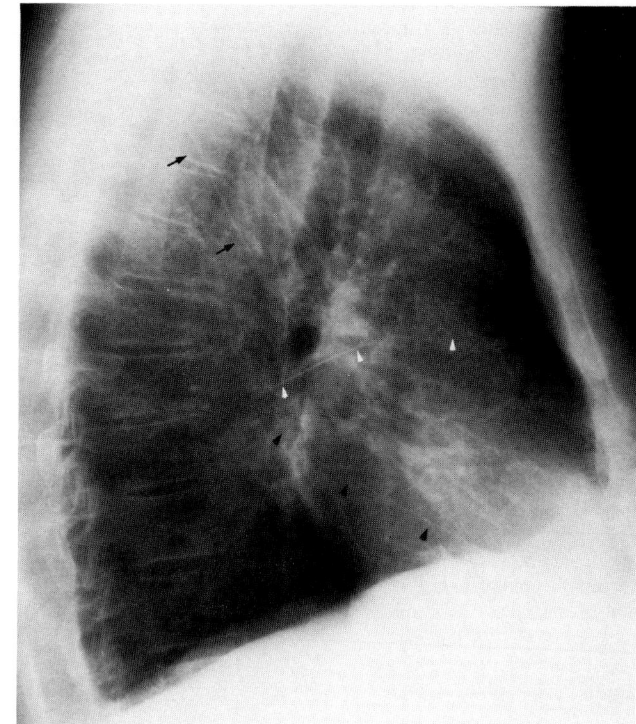

B

Fig. 2-1. *A.* Normal chest with minor fissure outlined by arrows. *B.* White arrowheads point to the minor fissure; black arrowheads, to the lower portion of the right major fissure; black arrows, to the upper portion of the left major fissure.

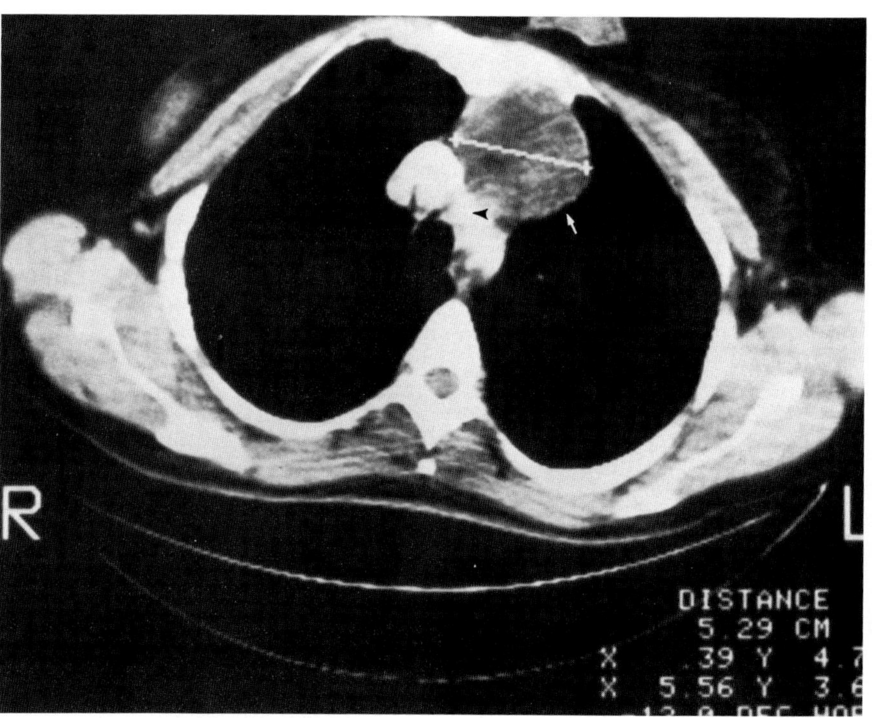

B

Fig. 2-2. *A.* Mediastinal mass (arrow) that has obliterated the outline of the main pulmonary artery segment and the cardiac border but not the outline of the descending aorta (arrowhead). This places the mass in the anterior mediastinum. *B.* The mass (arrow) anterior to the arch of aorta (arrowhead). *C.* Outline of the posteriorly located aortic knob and proximal descending aorta (arrows) is not obliterated by the anteriorly located mass. The mass did not obliterate hilar vessels that are clearly seen through it suggesting that it is more anterior to the hilar vessels. *D.* Arrows outline the pulmonary vessels seen through the mass. The fact that the hilar vessels are more than 1 cm medial to the edge of the mass suggests that it is not intrapericardial in location. This sign is called *hilum overlay.*

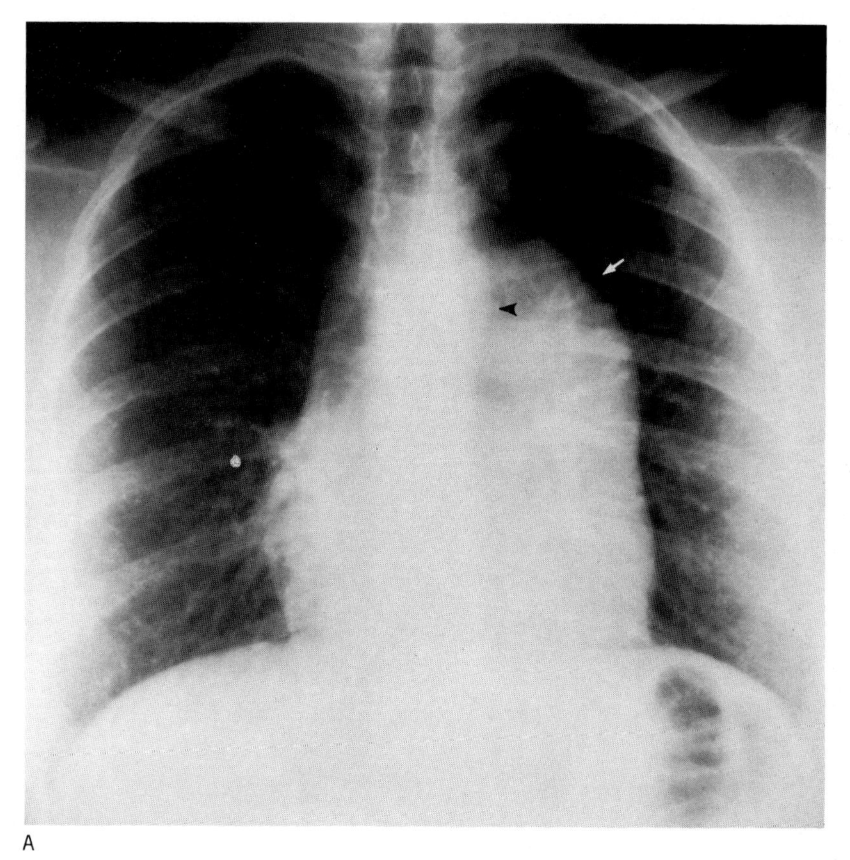

A

C

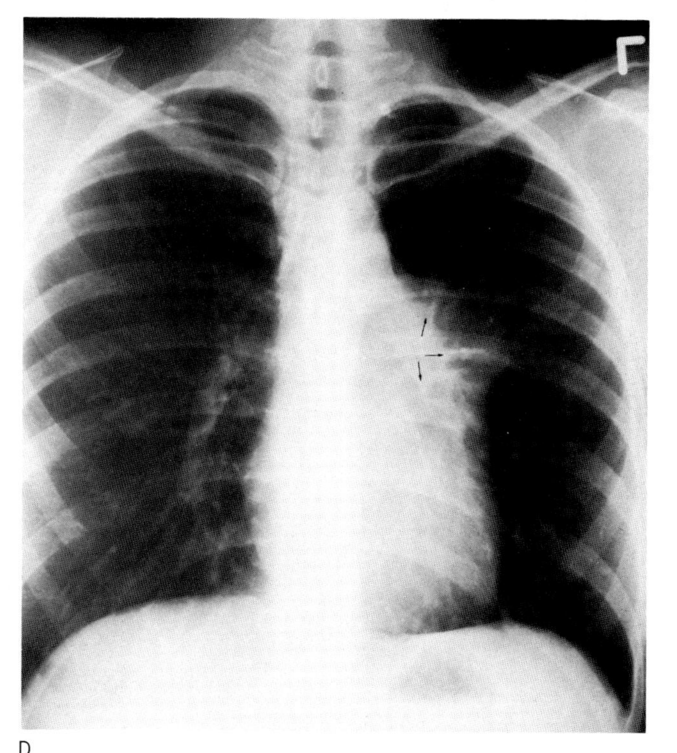

D

Fig. 2-3. Right paratracheal mass (arrow) obliterating the outline of the superior vena cava and ascending aorta.

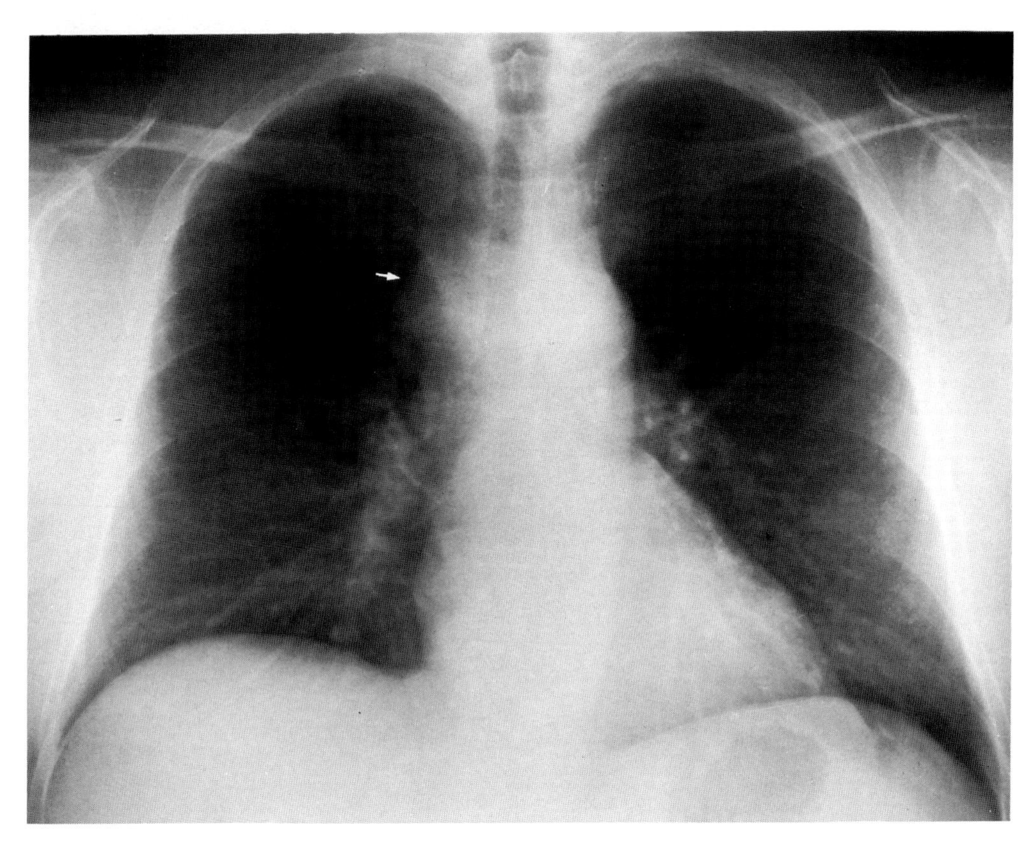

Fig. 2-4. *A.* A patch of pneumonia obliterating the right cardiac border. *B.* Lateral view of the patient in *A.* Arrows point to the right middle lobe pneumonia. *C.* Right middle and lower lobe atelectasis. The right cardiac border (arrow) and the outline of the right hemidiaphragm (arrowhead) have both been obliterated by the atelectatic lung. Compensatory hyperaeration of the right upper lobe with downward displacement of the major fissure has occurred.

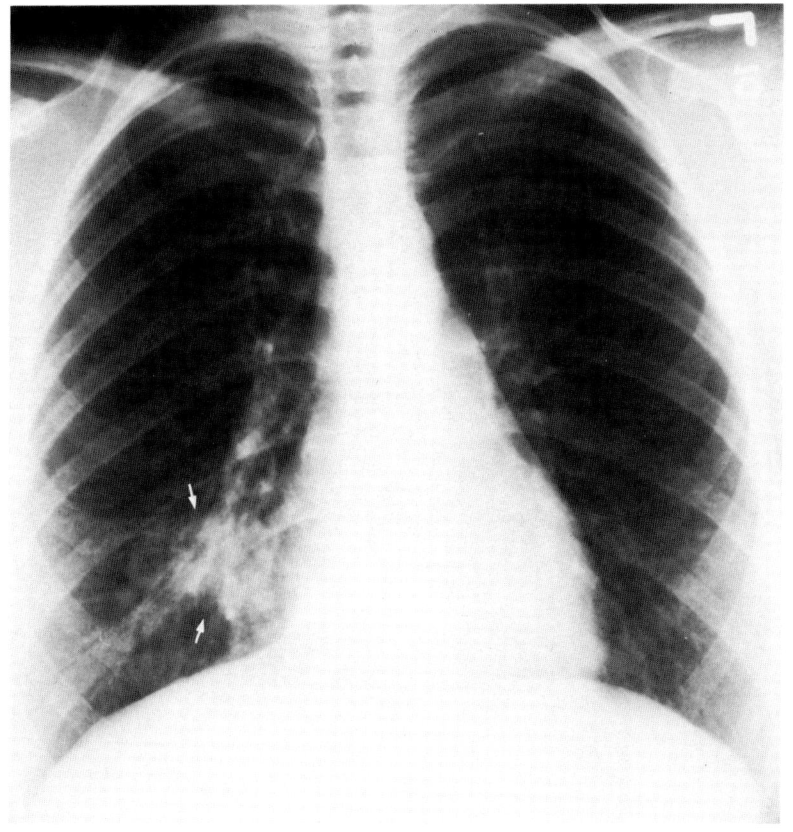

A

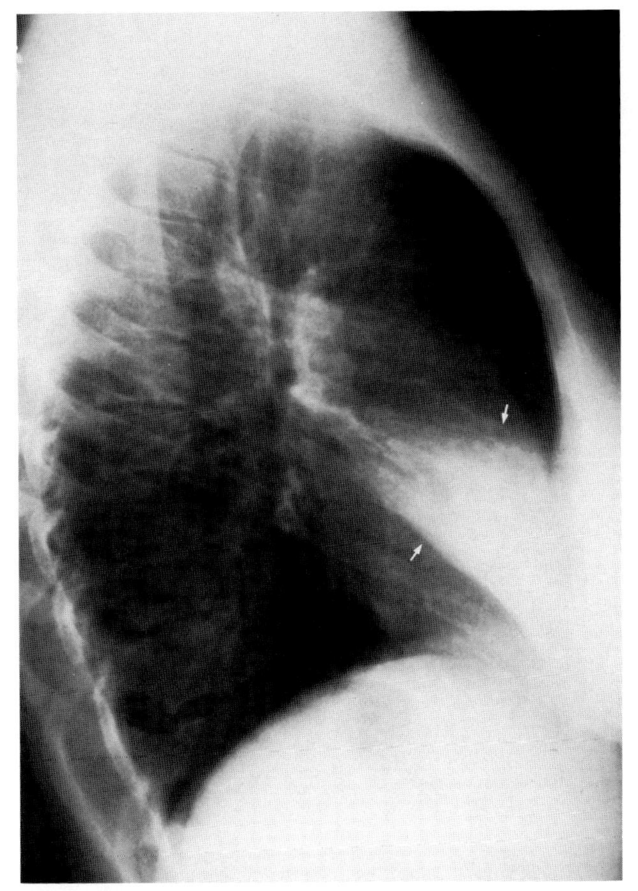

B

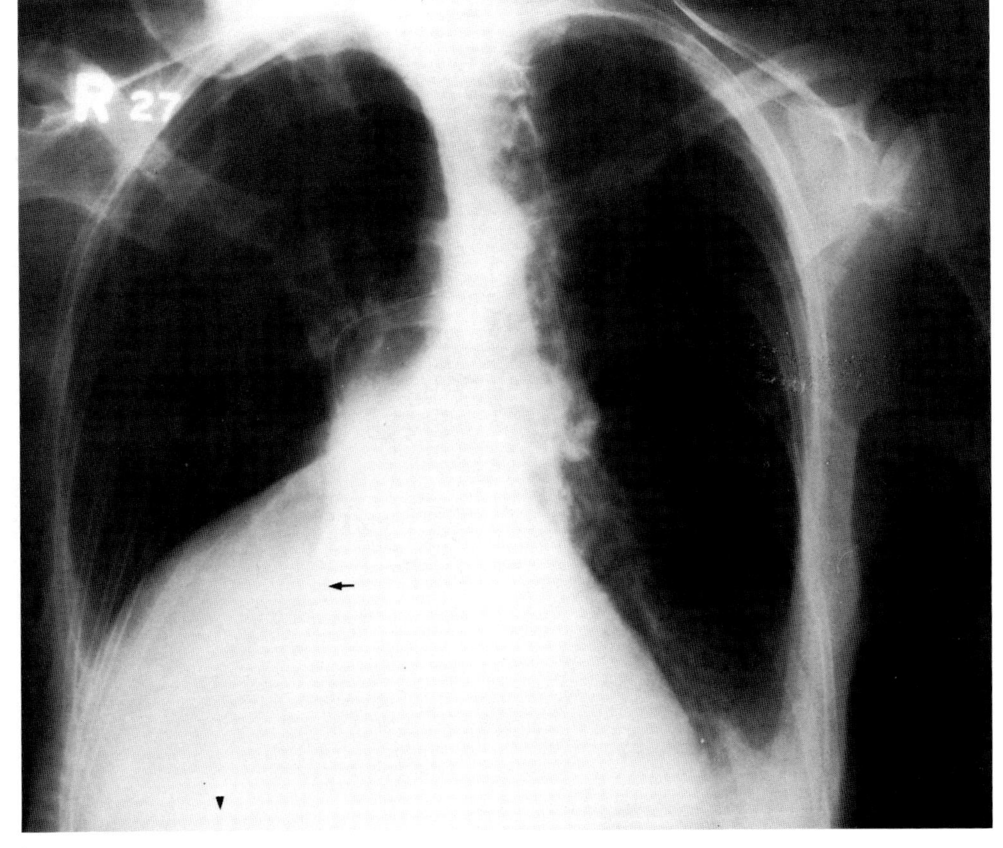

C

Fig. 2-5. *A.* Arrowhead points to a well-defined outline of a mass above the clavicle suggesting a posterior location. Arrow points to inferior margin of the mass disappearing or not well outlined above the diaphragm suggesting abdominal extension. *B.* Lateral of patient in *A* shows posterior location of the dilated esophagus displacing the trachea anteriorly (arrow) and obliterating the posterior cardiac border. Esophagus extends beyond thoracic cavity as it goes through the esophageal hiatus into the abdomen.

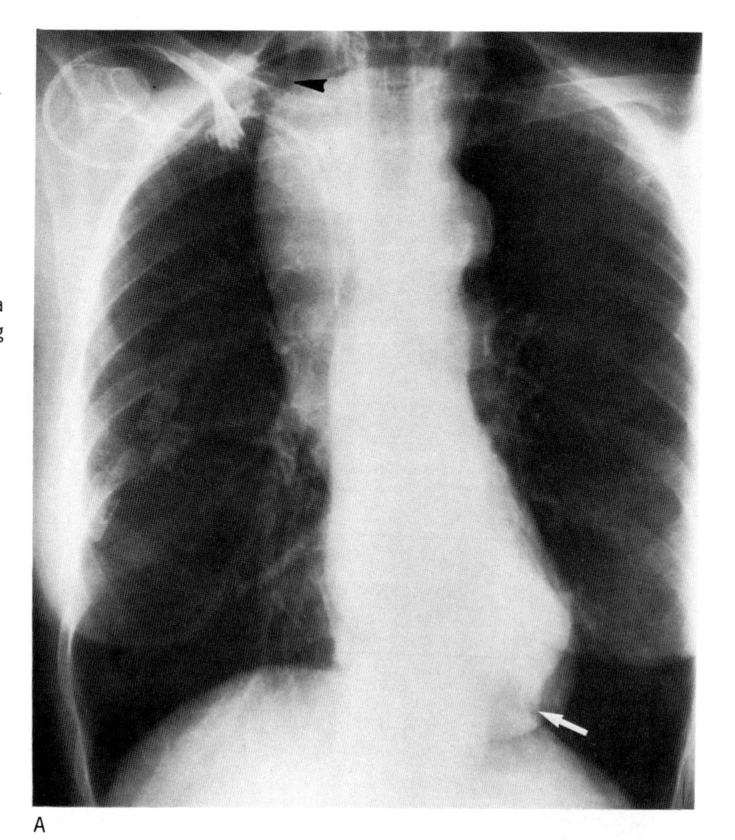

A

B

Fig. 2-6. Visible solitary pulmonary nodule (arrow). No obstructing vessels or structures obscure the lesion. Arrowheads point to infiltrates in the retrocardiac region.

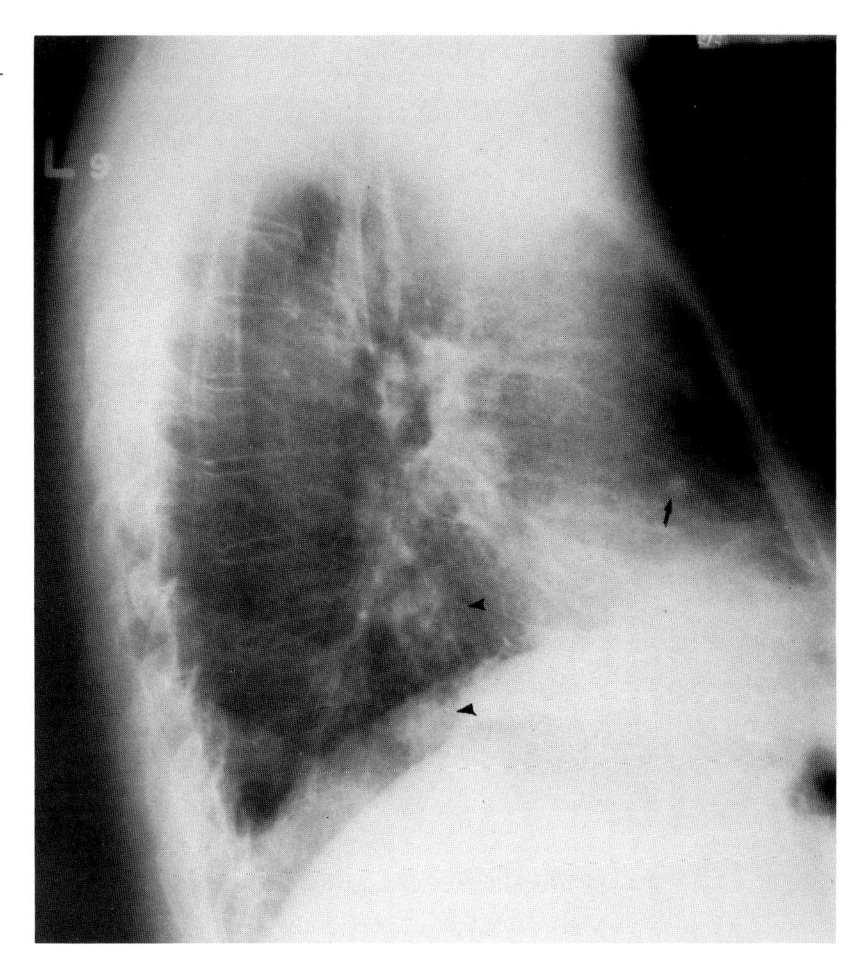

through the diaphragmatic shadow, it is totally intrathoracic; if its lower border is obliterated by the diaphragm, it either ends at the diaphragm or extends below it.

Radiologically, it has been shown that the threshold of visibility of a density ranges from 3 to 6 mm for solitary lesions [3,4]. These lesions are visible if located in the interspaces and unobstructed by overlying ribs or vessels (Fig. 2-6). These lesions may be missed if located in the pleura or hilar areas; over the mediastinum, hila, or diaphragm; or in or under the pleura.

Multiple lesions are subject to the effect of summation. There are various opinions on this subject. Some authors claim that the loss of visibility of individual images is secondary to superimposition of shadows [5], whereas others claim that lesions are raised to the threshold of visibility by a summation of shadows [6].

CONSOLIDATION

Consolidation refers to the filling of acini with fluid, tissue, or exudate. (An acinus is a unit of lung from the terminal bronchiole and all the lung subtended by it.) Consolidated areas vary in size from 7 mm (the size of an acinus) to the entire lung. The following features are helpful in determining the nature of the consolidation:

1. Segmental or nonsegmental distribution. Widely disseminated diseases such as pulmonary edema and acute alveolar pneumonias (except pneumococcal and *Klebsiella* pneumonias) usually show no definite segmental distribution (Fig. 2-7), whereas pulmonary infarcts (Fig. 2-8) and bronchogenic carcinomas do.
2. Margins of lesions. Margins are either smooth and well circumscribed or poorly defined. Metastatic nodules, lesions abutting a fissure, and intrafissural lesions show smooth, well-defined borders (Fig. 2-9).
3. Presence or absence of an air bronchogram. When the amount of air in the alveoli has been reduced significantly because it has been replaced or absorbed, but the supplying bronchus has remained patent, an air bronchogram is seen [7,8,9]. Pneumonic consolidations, lymphoma, pseudolymphoma, and respiratory distress syndrome (both in adults and in the neonate) may show air bronchogram (Fig. 2-10).

When we speak of consolidation, we most often use the term to mean an abnormality due to a pneumonic process. Pneumonia, however, may not always present as a segmental or lobar consolidation. It may show a bronchopneumonic pattern or an interstitial pattern. Several authors [10,11,12] have shown that it is impossible to predict accurately the etiologic agent (i.e., to determine whether it is bacterial or viral) on the basis of the radiologic appearance of the pneumonic process. Acute interstitial pneumonia may produce sufficiently large amounts of alveolar fluid to simulate a lobar pneumonia. A bronchopneumonic process or a lobular process may be so confluent as to appear radiologically as a lobar consolidation. However, there are some typical patterns produced by certain organisms. Table 2-1 lists these characteristic patterns.

Genereux and Stilwell [13] aptly summed up the indications for doing a chest x-ray examination in a patient suspected of having a pneumonia, knowing full well that the etiologic agent cannot be accurately predicted:

TABLE 2-1. Characteristic Patterns of Parenchymal Involvement and Associated Organisms

Pattern of parenchymal involvement	Organism
Lobar consolidation (Fig. 2-11)	*Streptococcus pneumoniae* (pneumococcus)
Bronchopneumonia (Fig. 2-12)	*Staphylococcus aureus* *Klebsiella* *Escherichia coli* *Pseudomonas* *Haemophilus influenzae* *Legionella pneumophilia* *Proteus* *Serratia* *Bacteriodes*
Interstitial infiltrates	Viral Mycoplasma
Bulging fissures (Fig. 2-13)	*Klebsiella* *Streptococcus pneumoniae*
Cavity formation (Fig. 2-14)	*Staphylococcus* *Klebsiella* *Streptococcus* *Pseudomonas* *Legionella pneumophilia*
Reticulonodular interstitial infiltrates	Psittacosis Mycoplasma in a patient with underlying lung disease or previous mycoplasma infection

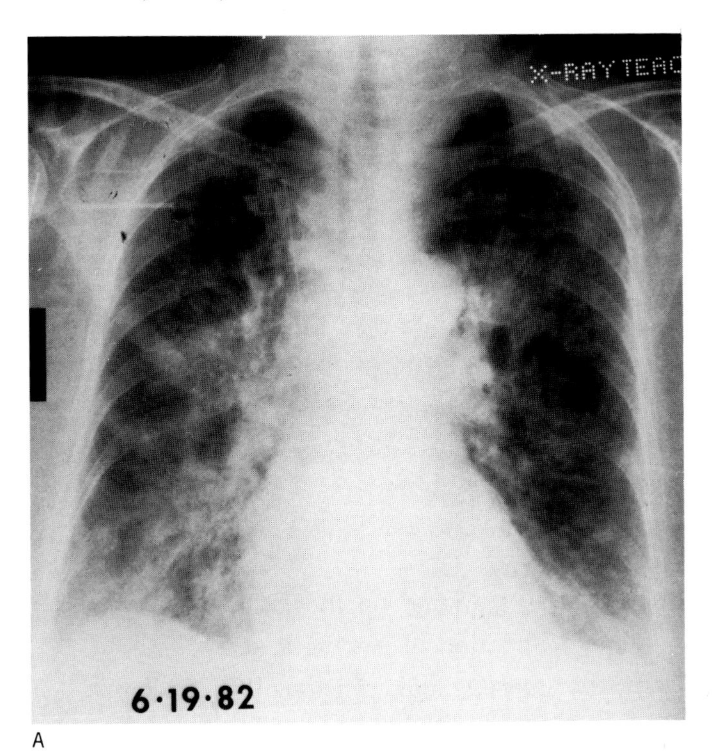

A

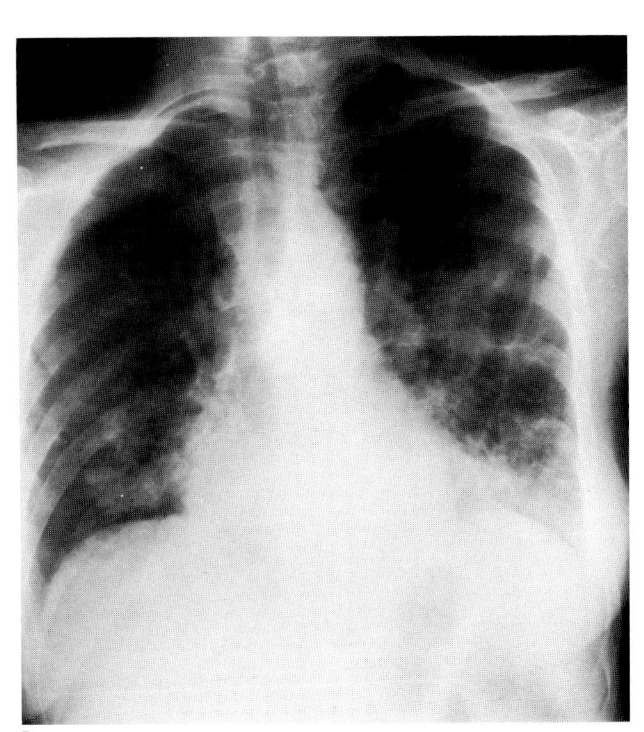
B

Fig. 2-7. *A.* Nonsegmental distribution of densities in a patient with *Pneumocystis carinii* pneumonia. *B.* Nonsegmental distribution of infiltrates in a patient with aspiration pneumonia.

Fig. 2-8. *A.* Lung infarct (arrow). Note segmental distribution of the lesion. Infarct was caused by too distal positioning of Swan-Ganz catheter. *B.* Patient in *A* 1 year later. Partial resolution of the infarct still maintains its shape but has decreased in size.

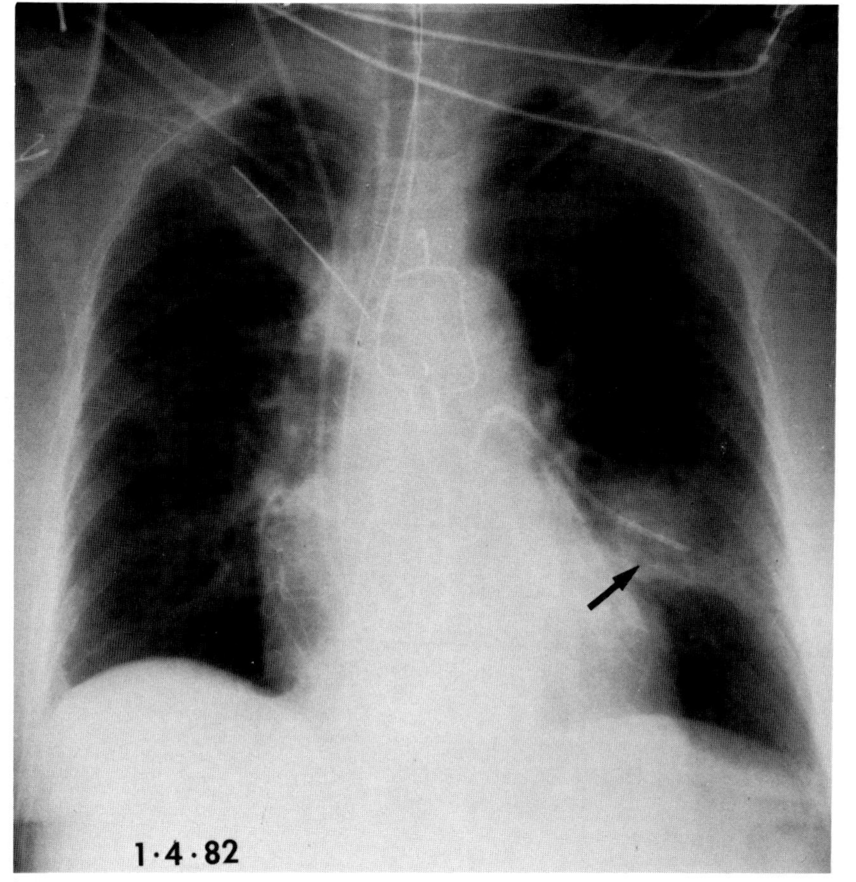

A

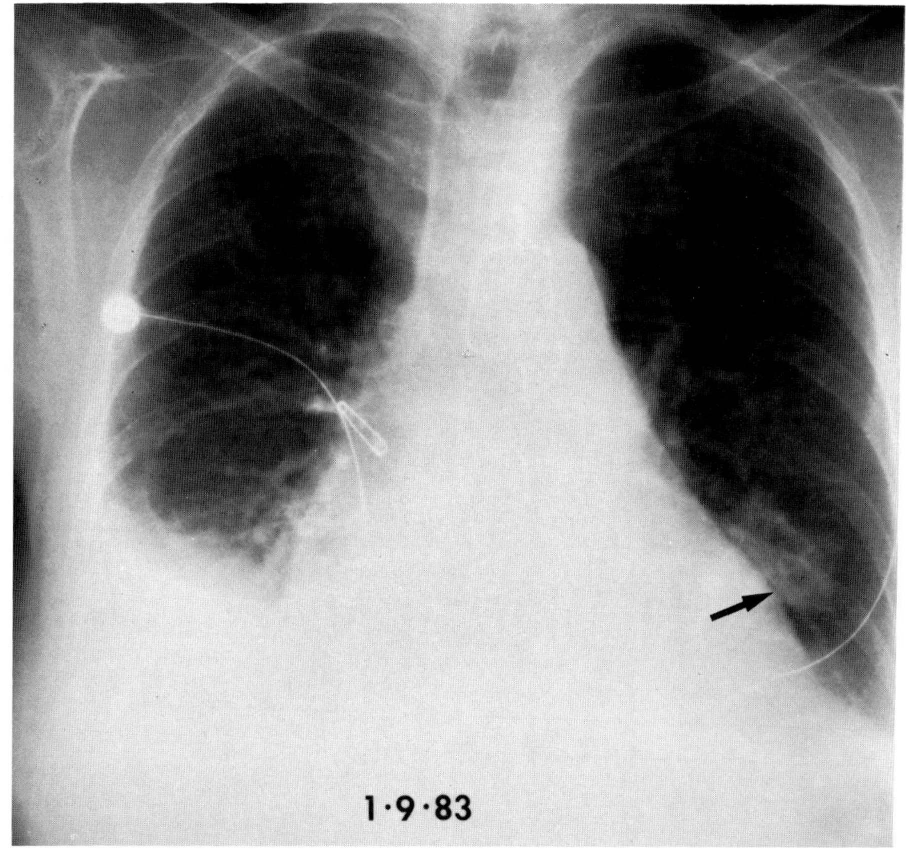

B

Fig. 2-9. *A.* Multiple metastatic nodules with well-defined borders. *B.* Fluid within minor fissure (arrow) shows well-defined borders.

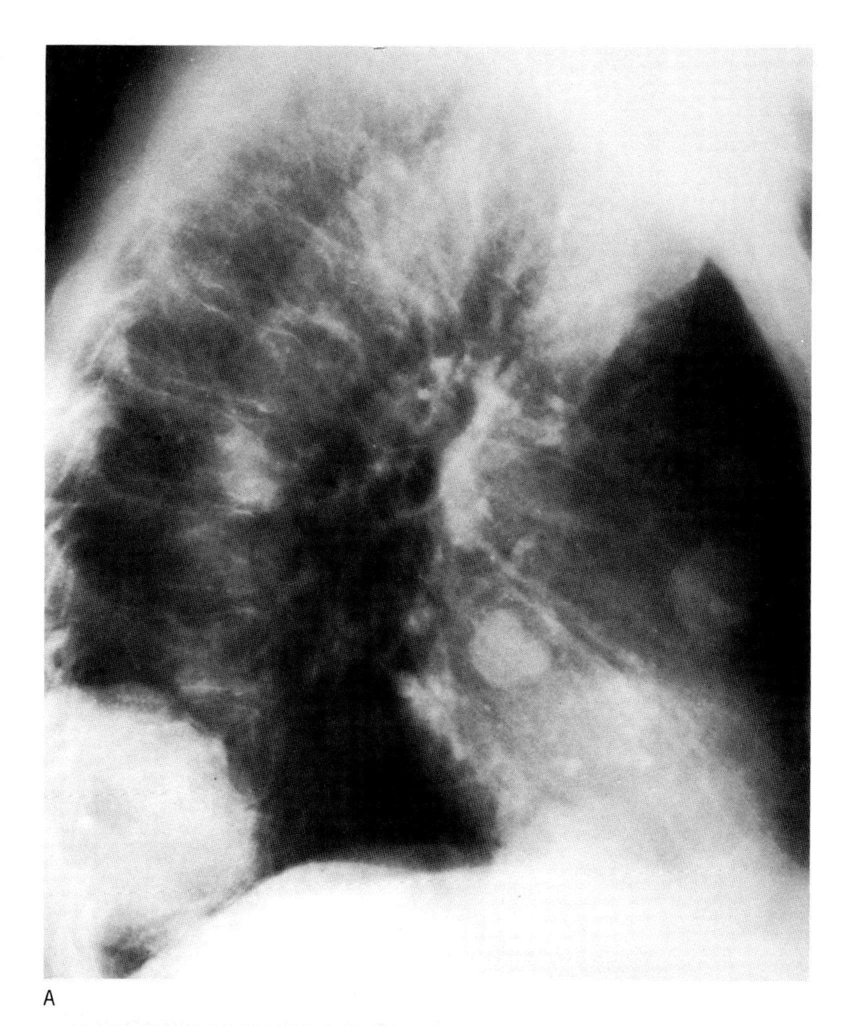

A

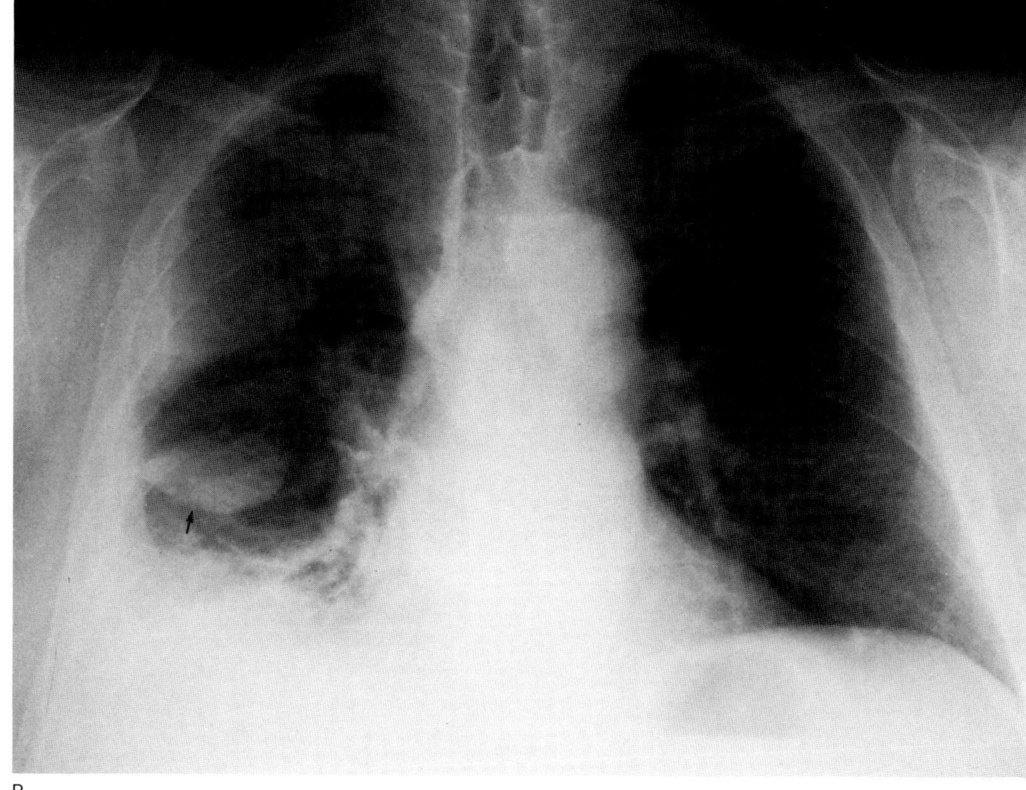

B

Fig. 2-10. *A.* Air bronchograms (arrows) in a patient with adult respiratory distress syndrome. *B.* Air bronchograms (arrows) within a consolidated lobe from pneumococcal pneumonia.

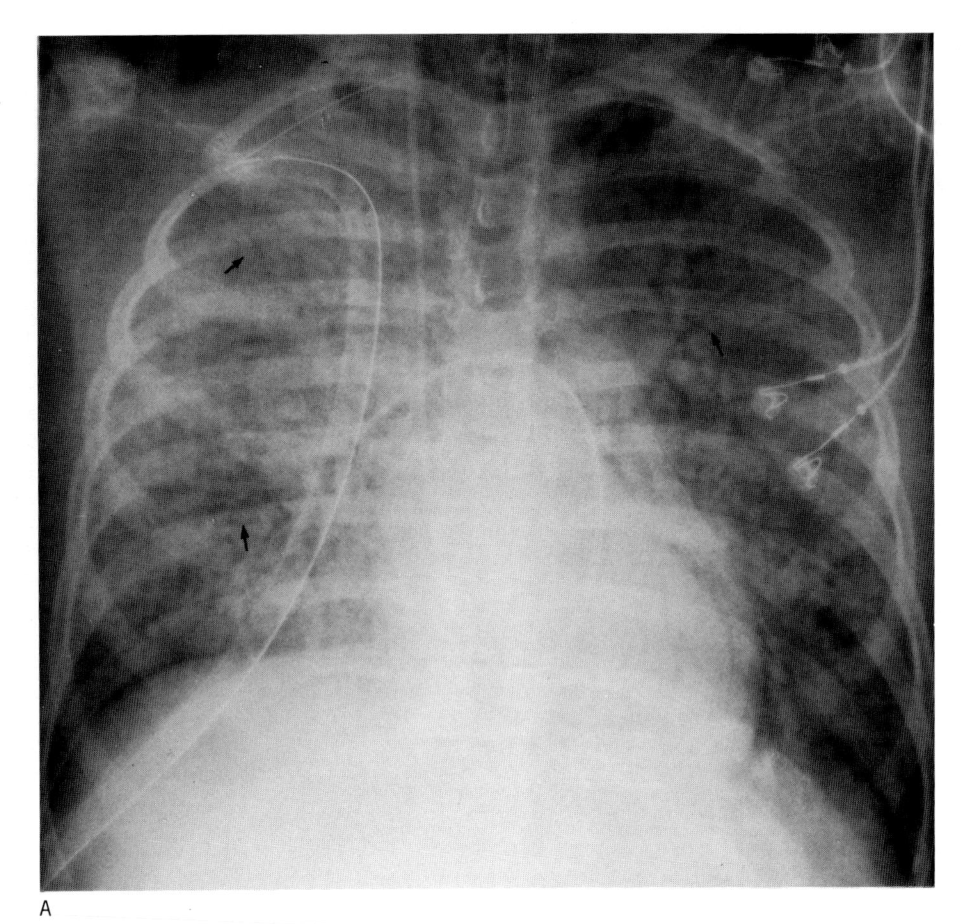

A

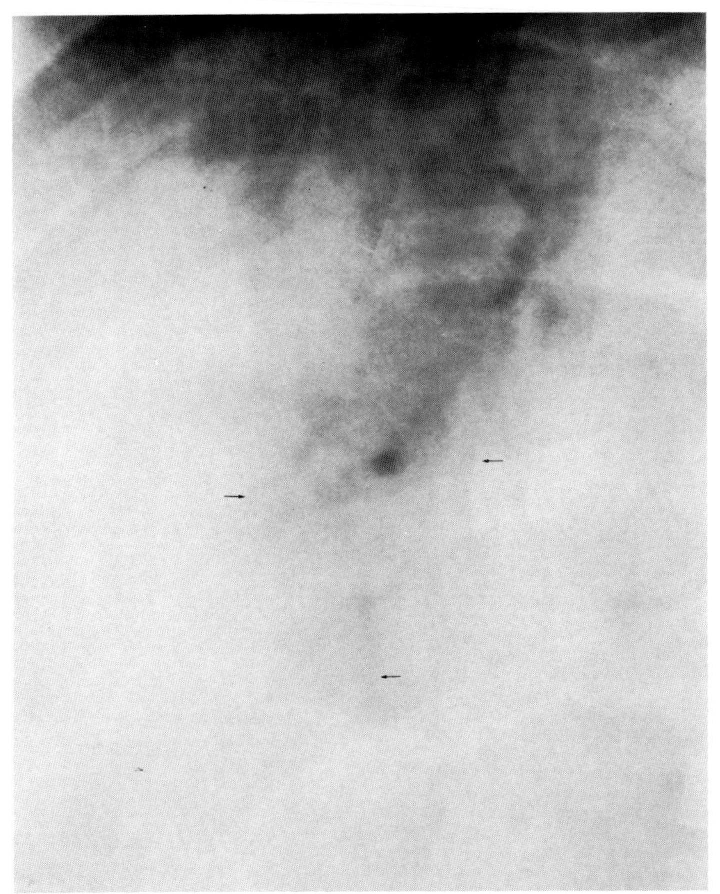

B

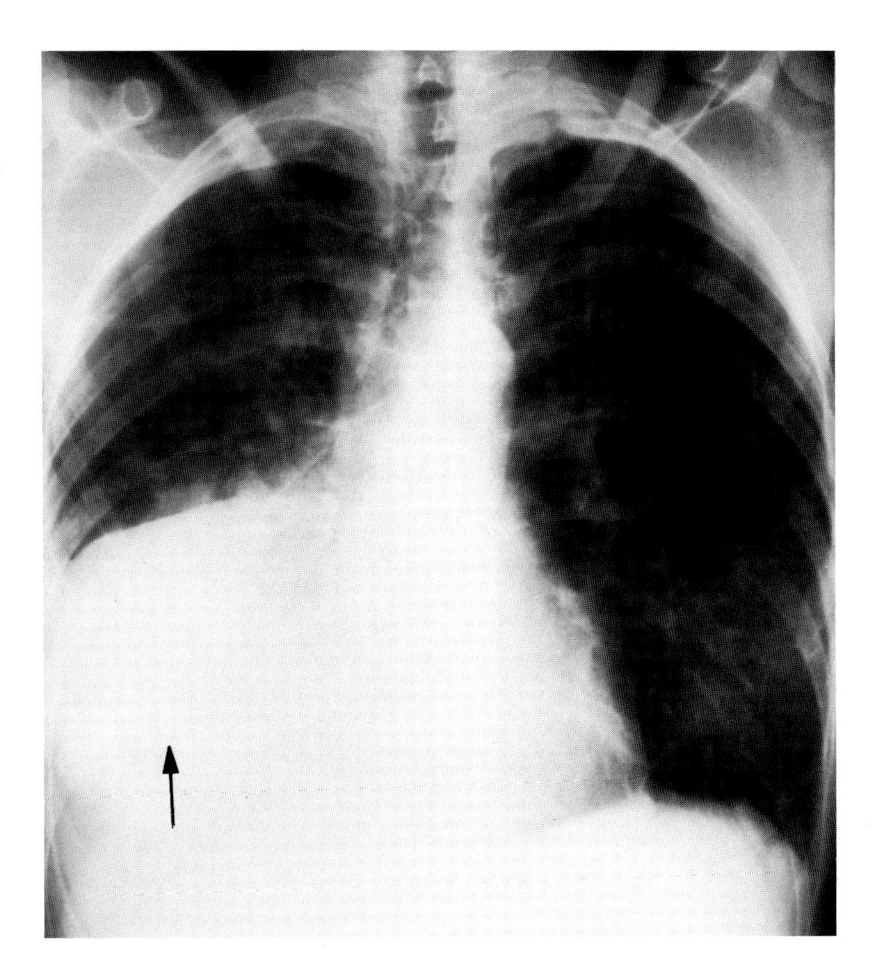

Fig. 2-11. Lobar consolidation: Right middle and lower lobe consolidation (arrow) with air bronchograms from pneumococcal pneumonia (same patient as in Fig. 2-10*B*).

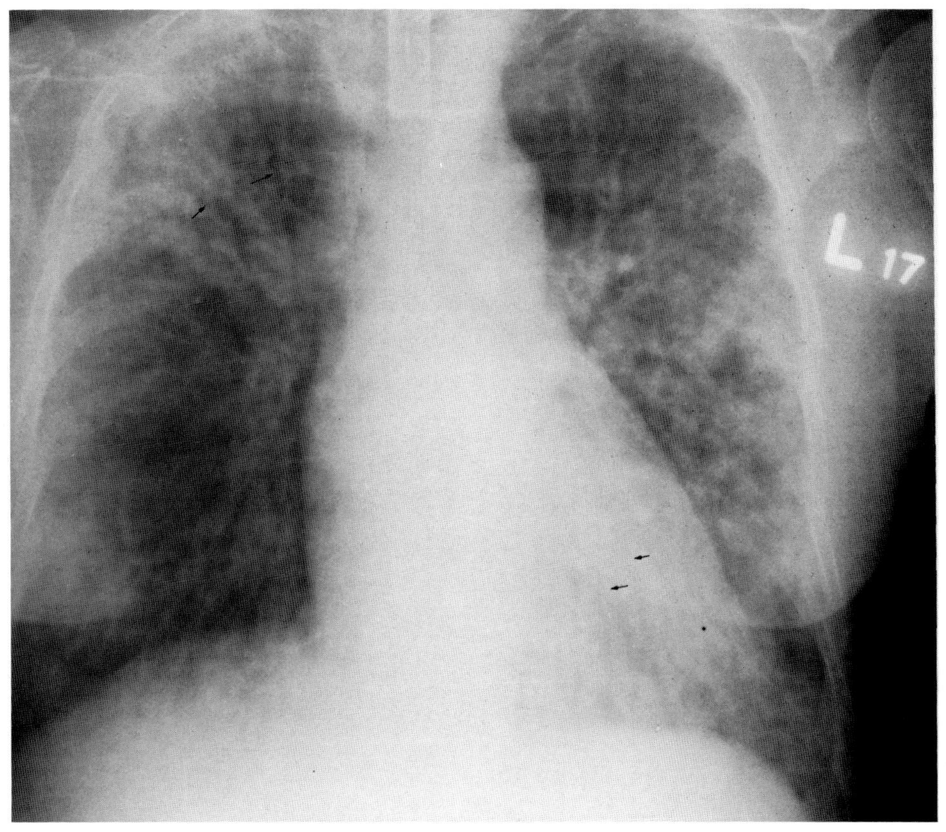

Fig. 2-12. The lucent tubular structures (arrows), representing air-filled bronchi (air bronchograms), in a patient with staphylococcal bronchopneumonia.

1. To determine if a radiologically detectable abnormality exists.
2. To determine if the visualized abnormality is compatible with a pneumonic process.
3. To determine the location and extent of pathology.
4. To assess the evolution of the process.
5. To identify complications.

ATELECTASIS

Atelectasis denotes reduction in volume secondary to diminished alveolar air content due to absorption of air without its replacement. There are four types of atelectasis [14]:

1. Passive relaxation or compression atelectasis.
2. Cicatrization atelectasis.
3. Adhesive atelectasis.
4. Resorption atelectasis secondary to bronchial obstruction.

An air bronchogram may be visible in the first three types of atelectasis; however, it is not seen in the last type. In a healthy lung, all air is absorbed within 24 hours. If a person has been given oxygen prior to atelectasis, absorption of alveolar air is faster, probably within minutes [15], because oxygen is absorbed 60 times faster than air [16].

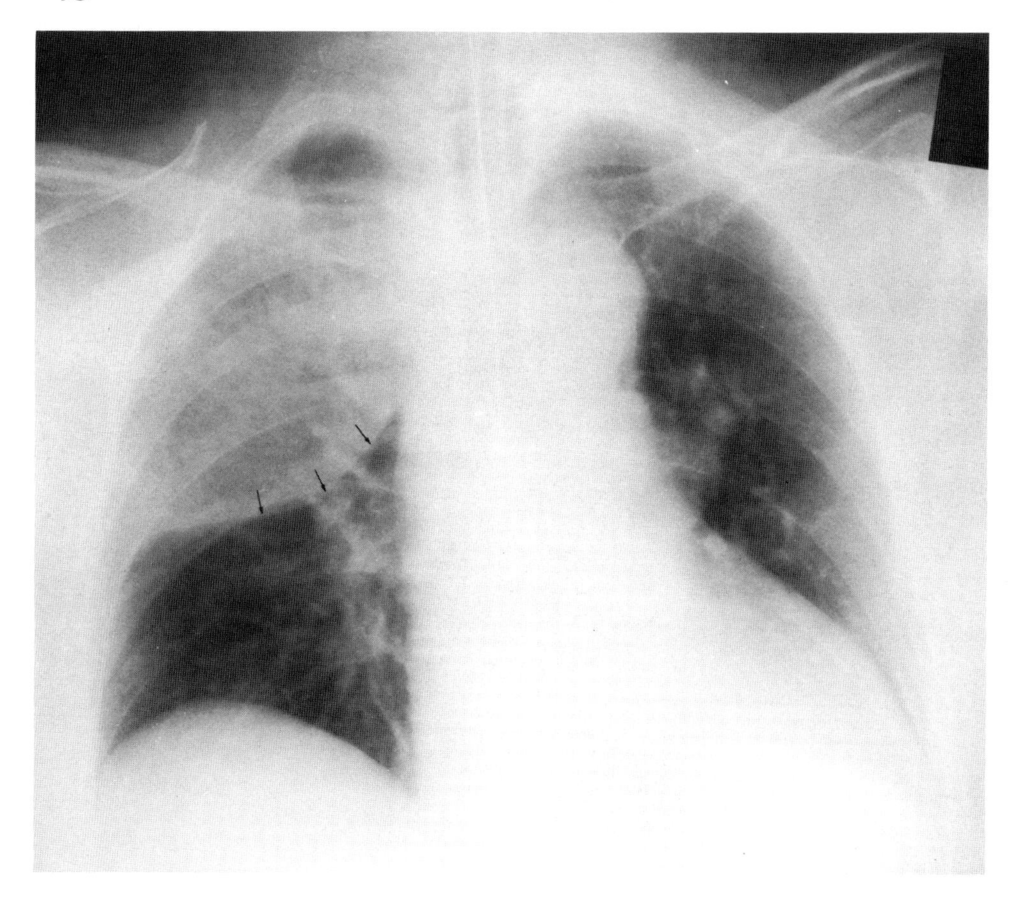

Fig. 2-13. Lobar consolidation from *Klebsiella* pneumonia. Arrows point to the bulging minor fissure.

Fig. 2-14. *A.* PA view of staphylococcal pneumonia with numerous abscesses. *B.* Lateral view of the patient in *A.* *C.* Arrowhead points to the cavity developing within an area of streptococcal pneumonic lobar consolidation (arrows).

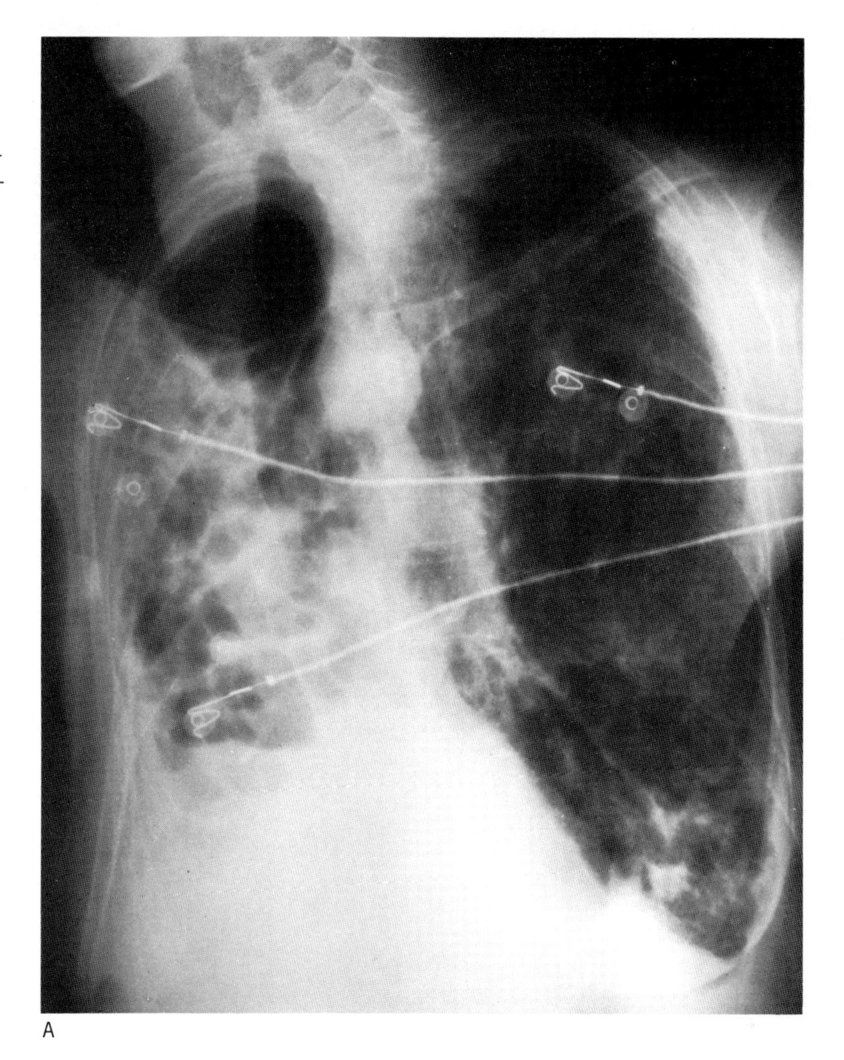

A

B

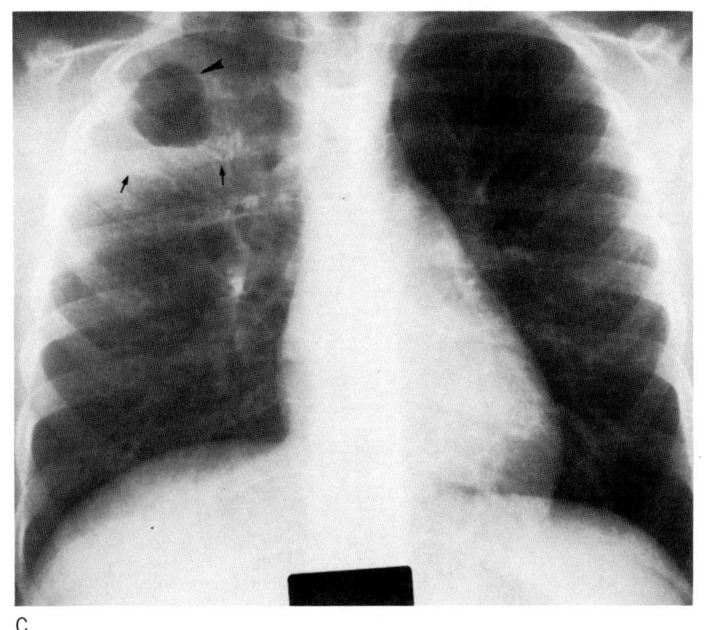

C

Fig. 2-15. Total collapse of the left lung. Note opaque left hemithorax, shift of the mediastinum to the left, and total cutoff of air column in the left main bronchus (arrow).

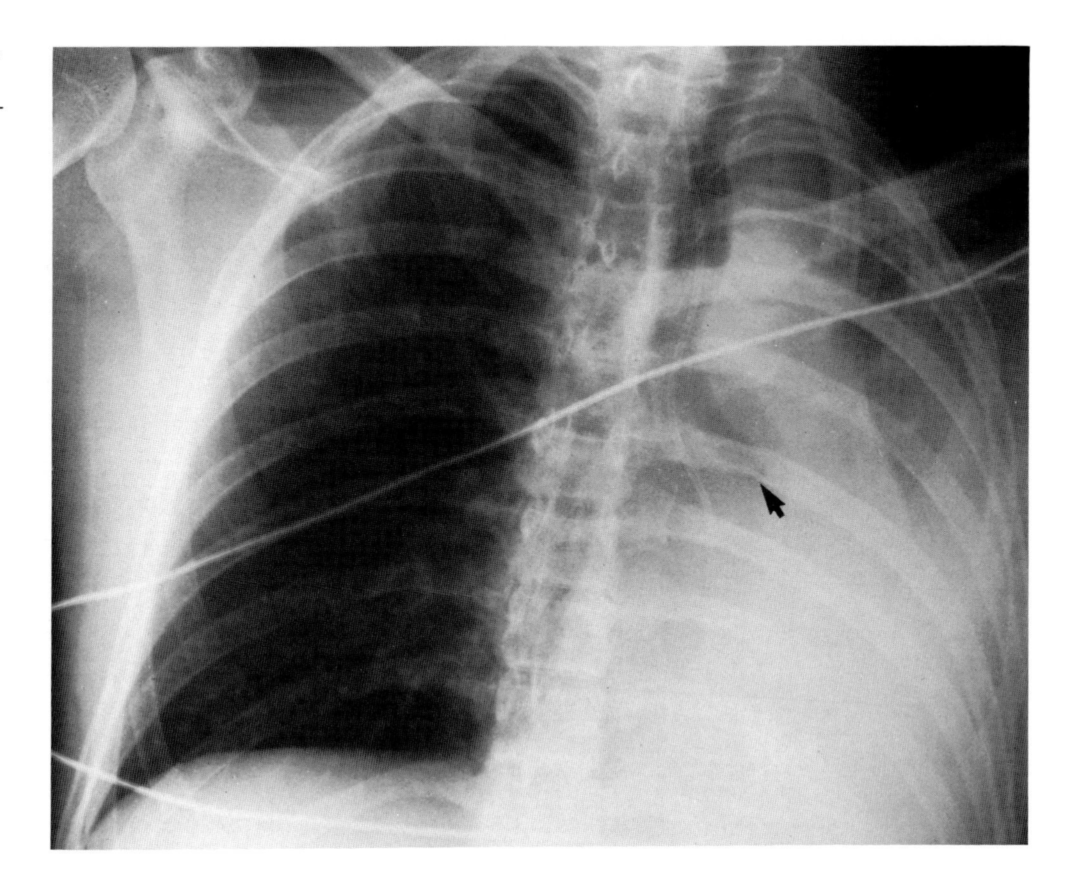

Obstructive atelectasis may be subdivided into large and small airway obstructions. Large airway obstruction is often secondary to the presence of a foreign body or endobronchial masses as seen in bronchogenic carcinoma, bronchial adenoma, endobronchial granuloma (tuberculosis, rarely sarcoid), papilloma, inflammatory polyps, rare cases of endobronchial metastasis, and rarer cases of benign tumors such as granulosa-cell myoblastoma and amyloidosis. Extrinsic airway compression can also cause large airway obstruction, such as in left atrial enlargement and vascular rings. Small airway obstruction is most often caused by mucous plugging. Occasionally, collateral ventilation through the pores of Kohn and canals of Lambert may prevent atelectasis, despite total bronchial occlusion.

Total Atelectasis Of the several typical patterns of pulmonary collapse, total atelectasis produces the greatest number of changes. In total pulmonary atelectasis of one side, there is a shift of the mediastinum to the side of collapse, the hemidiaphragm moves upward, and one sees in the lateral view an increase in the depth and lucency of the retrosternal air space associated with an increase in density of the posterior portion of the thorax, as the collapsed lung tends to rotate posteriorly (Fig. 2-15A). In the posteroanterior (PA) view, one sees a slight increase in the lucency of the opacified hemithorax if herniation of the opposite lung across the midline has occurred (Fig. 2-15B).

Of the many radiologic signs associated with significant atelectasis, the most direct is displacement of the interlobar fissures. Other signs are elevation of the hemidiaphragm, displacement of the mediastinum, hilar displacement, change in intercostal distances, and compensatory hyperaeration of the adjacent lung producing a localized decrease in lung density. Small areas of atelectasis may be indistinguishable radiologically from areas of consolidation, because these compensatory signs are not manifested. Although atypical patterns of collapse occasionally occur, discussion of those numerous variations will not be dealt with in this book. The typical appearances of the various forms of lobar collapse (Fig. 2-16) are described below.

Left Upper Lobe Collapse Left upper lobe collapse on the lateral view shows progressive anterior displacement of the major fissure, producing a density that almost parallels the sternum (Fig. 2-17). On the PA view, the collapsed lobe obliterates

Fig. 2-16. Diagrams of (*A*) left upper lobe collapse, (*B*) progressive right upper lobe collapse, (*C*) progressive right middle lobe collapse in the lateral view, and (*D*) progressive lower lobe collapse (same pattern of collapse for left and right lower lobes) in lateral and PA views. Sparsely dotted area represents minimal collapse; compactly dotted area, moderate collapse; lined area, severe collapse.

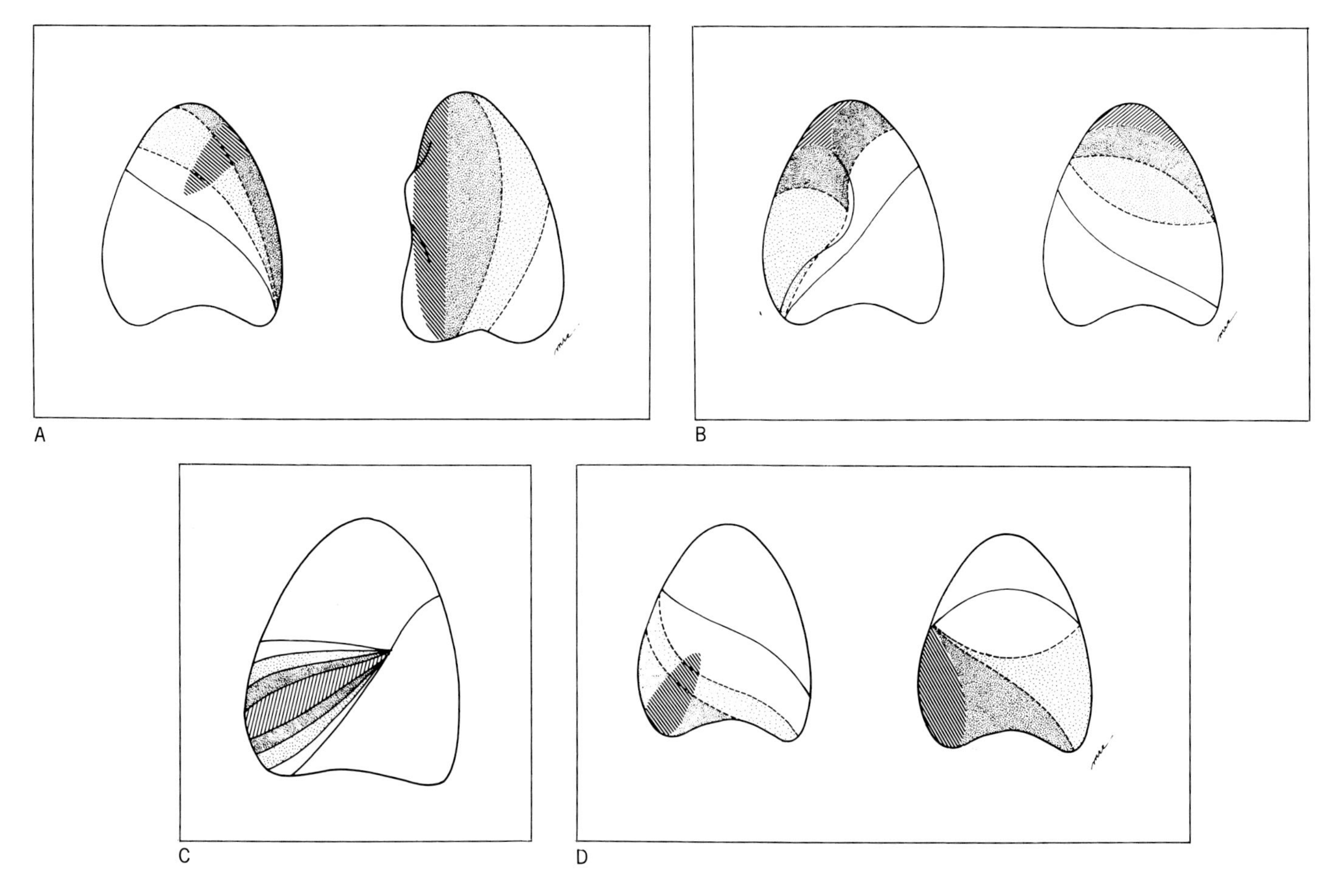

Fig. 2-17. *A.* Arrowheads outline the edge of the atelectatic left upper lobe. Arrow points to the left hilar mass, the cause of the atelectasis. *B.* Obliteration of the left cardiac border produced by left upper lobe atelectasis (arrowheads). Arrow points to a left hilar mass, the cause of the atelectasis. *C.* Lateral view of another patient with left upper lobe collapse (arrows). Note less straight, inferior edge of collapsed lung as compared with that in *A. D.* PA view of the patient in *C* showing a perihilar density and slight increase in upper lobe density as compared to the left lower lobe or right lung. *E.* Left hilar mass (arrow) producing left upper lobe atelectasis, the latter seen as an area of increased density (1) as compared to the base (2).

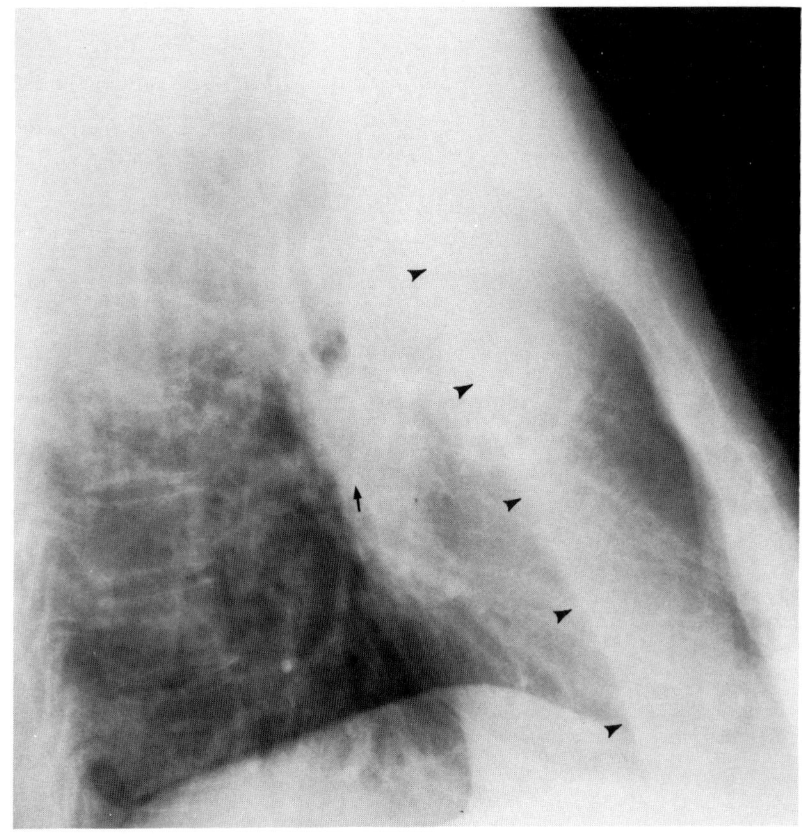

A

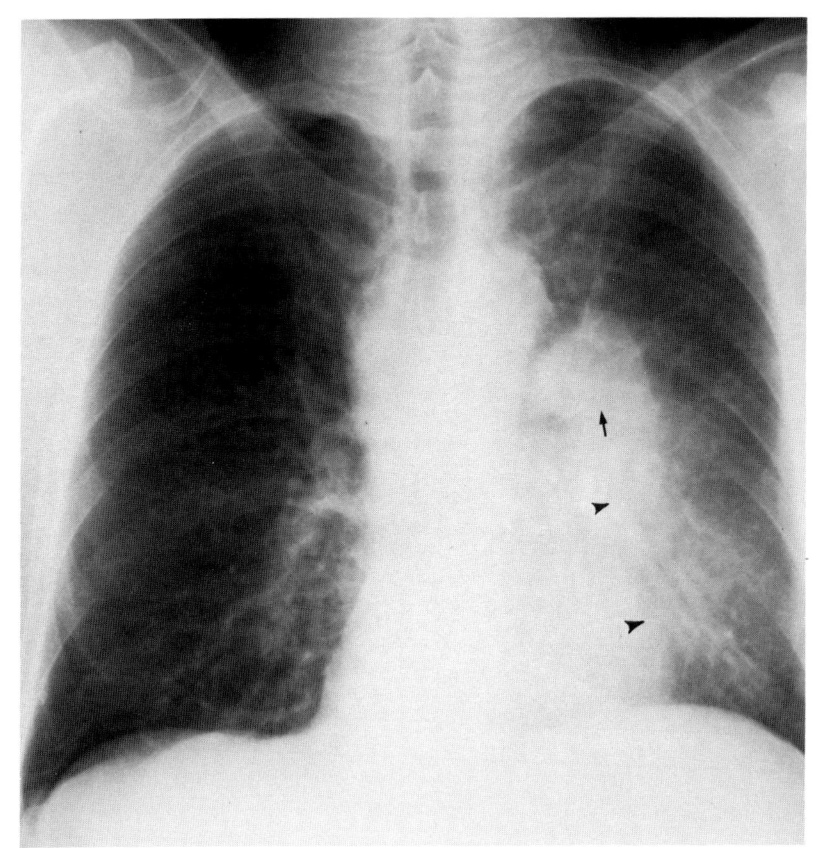

B

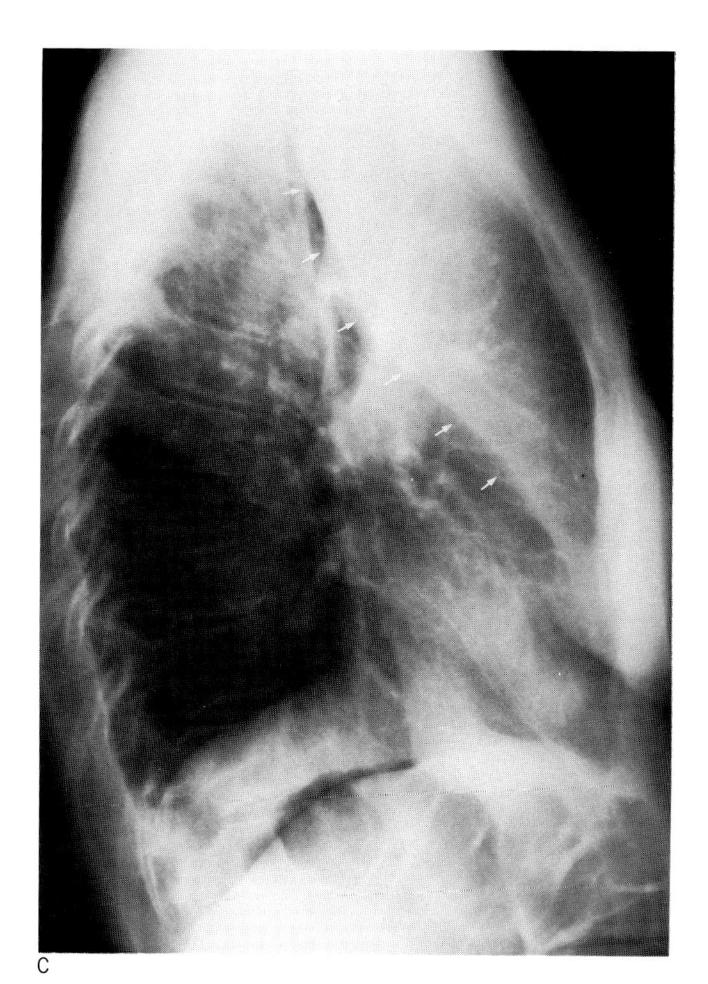

C

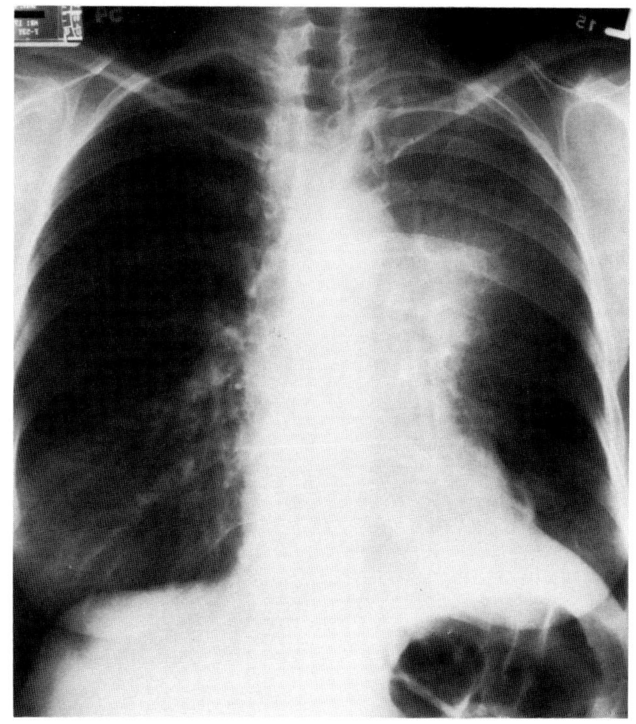

D

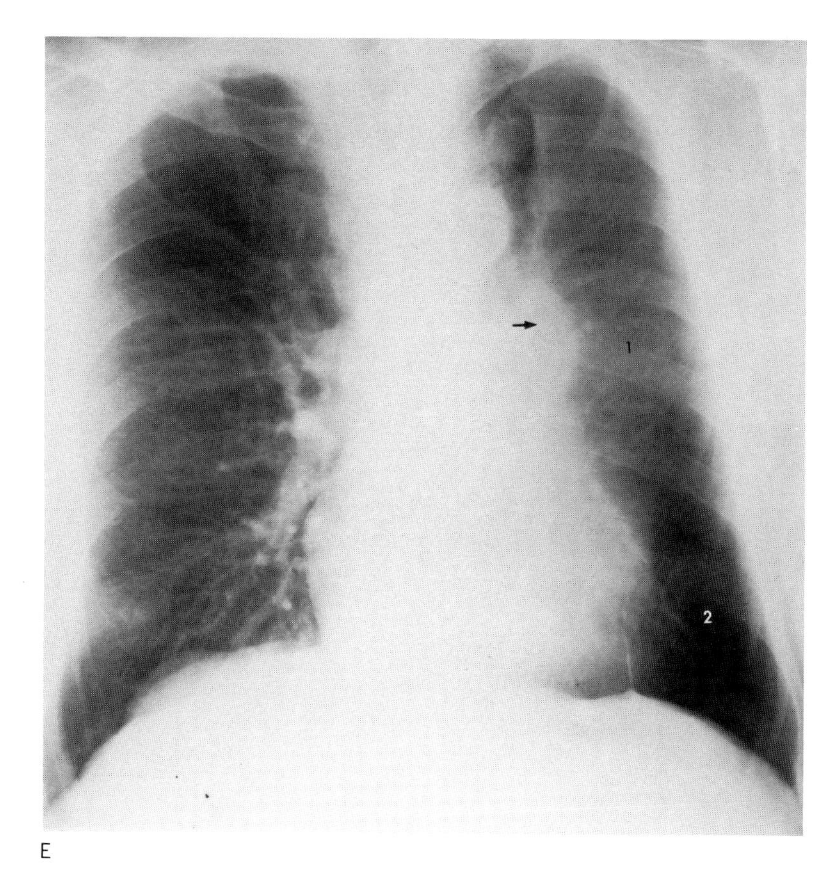

E

the left cardiac border. The left hilum becomes poorly defined, as the upper lobe artery is silhouetted out by the collapsed lung. A juxtaphrenic peak may be seen [17].

Right Upper Lobe Collapse The pattern of right upper lobe atelectasis is different from left upper lobe atelectasis because of the presence of the minor fissure. In both the PA and the lateral projections, right upper lobe collapse shows elevation of the minor fissure (Fig. 2-18) that varies from mild bowing or formation of the edge of a triangular density in the upper lobe to blending of the minor fissure with the superior mediastinum. The right hilum is elevated and diminished in size, and a hyperlucent lung with diminished vascularity is evident. A juxtaphrenic peak may be seen [17].

Right Middle Lobe Collapse Right middle lobe collapse is seen as obliteration of the right heart border on the PA view and downward displacement of the minor fissure (Fig. 2-19A). On the lateral view, it is seen as a wedge of triangular density with its apex directed toward the hilum (Fig. 2-19B). A lordotic projection shows the collapse best as a wedge with its apex pointed laterally (Fig. 2-19C).

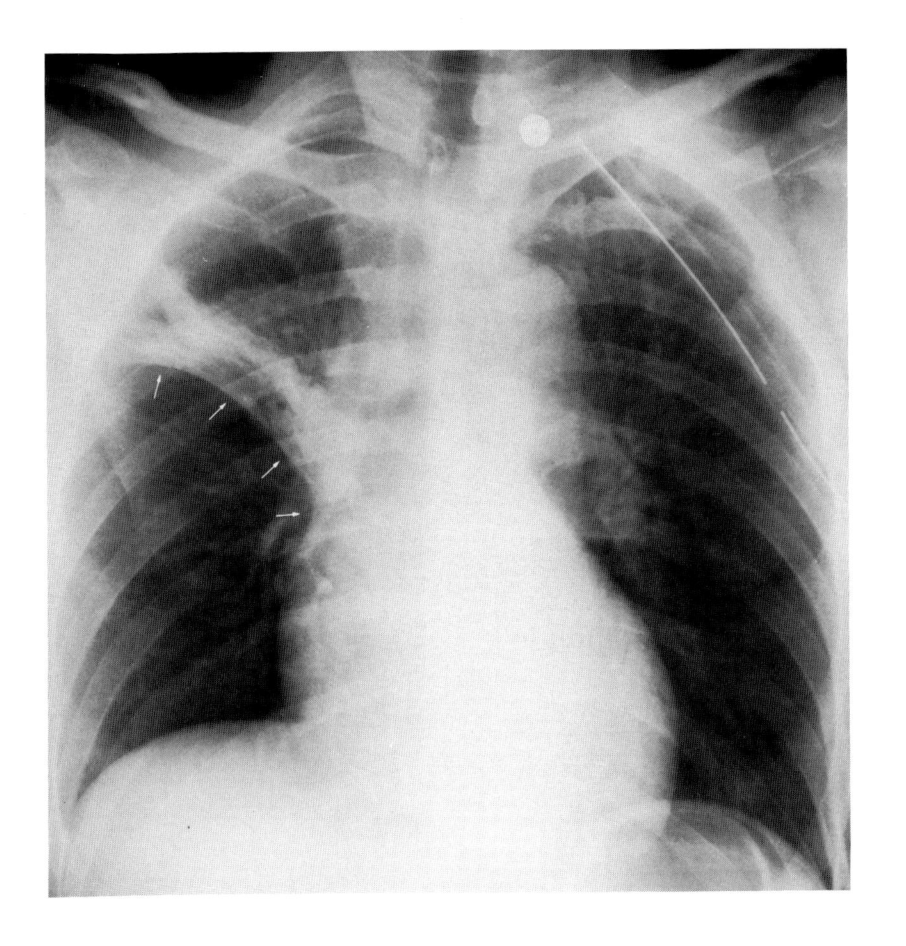

Fig. 2-18. Partial right upper lobe collapse. Note appearance of displaced minor fissure (arrows).

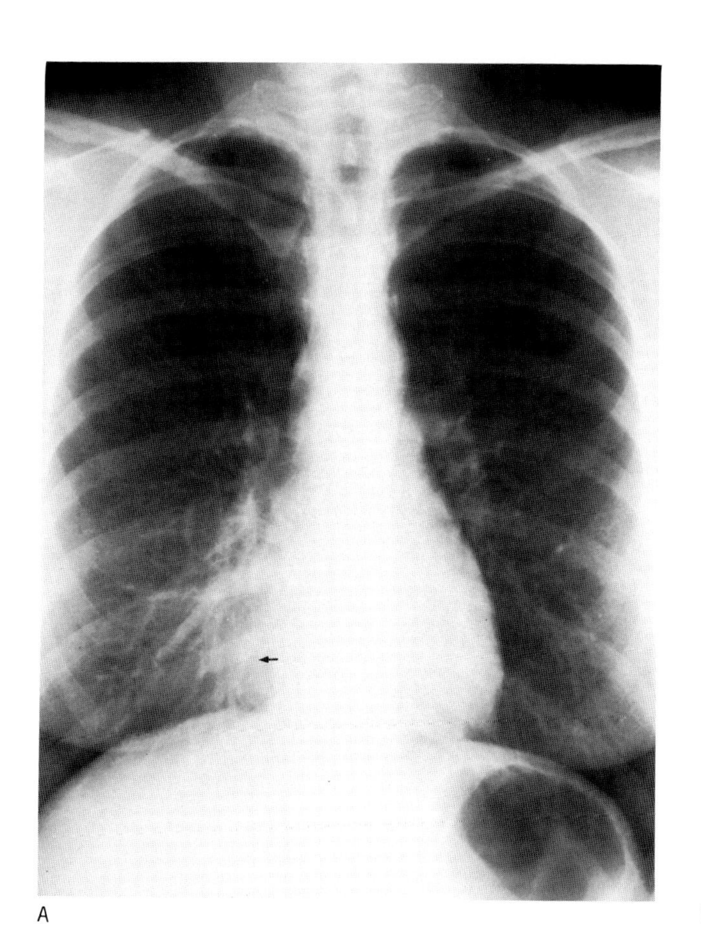

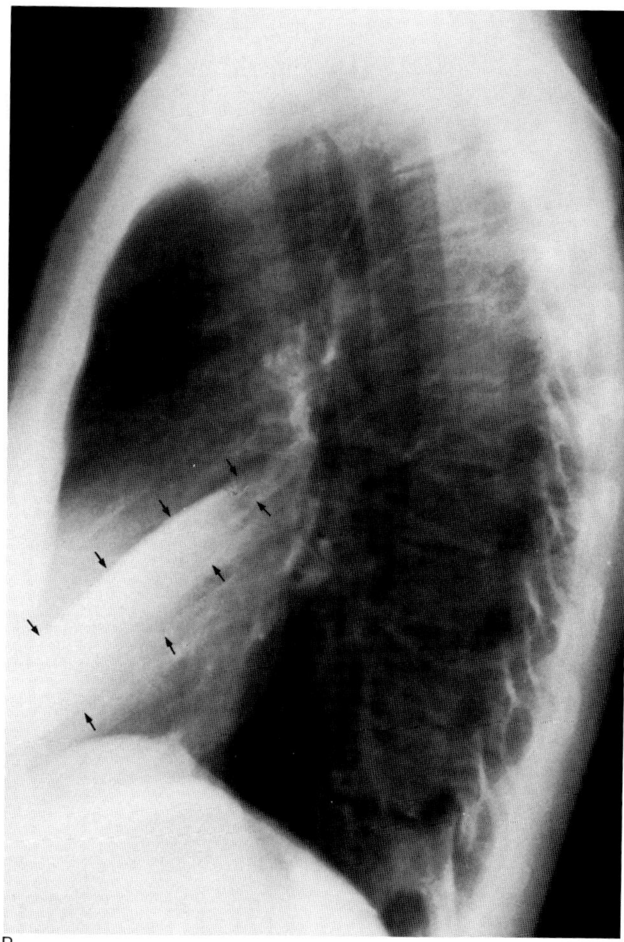

Fig. 2-19. *A.* PA view shows obliteration of the right cardiac border (arrow). *B.* Lateral view shows the right middle lobe atelectasis (arrows). *C.* Lordotic view shows the right middle lobe atelectasis (arrows).

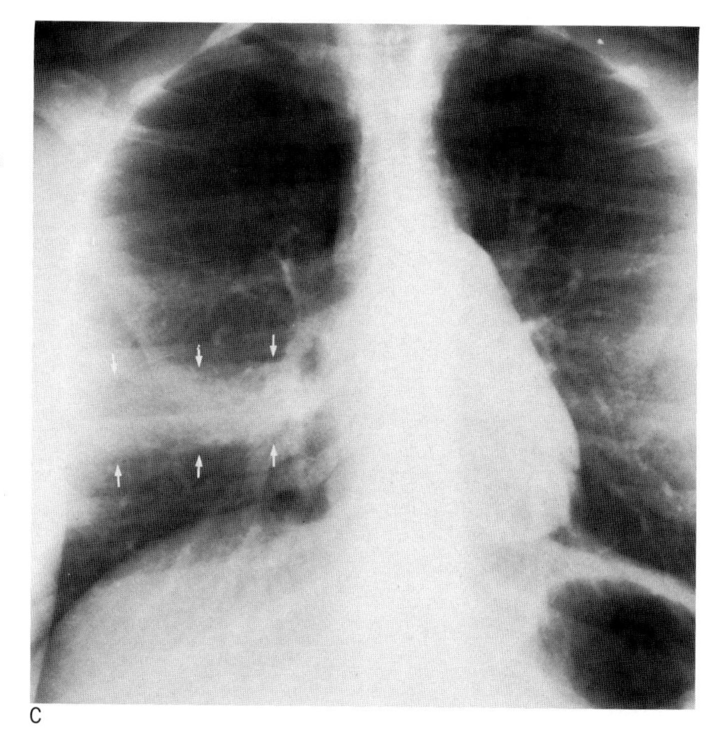

Lower Lobe Collapse Lower lobe collapse appears the same on the right and left sides. The lower lobe collapses toward the spine on the PA and lateral views, maintaining a wedge of dense tissue connected to the hilum by the inferior pulmonary ligament (Fig. 2-20). The major fissure is displaced, and

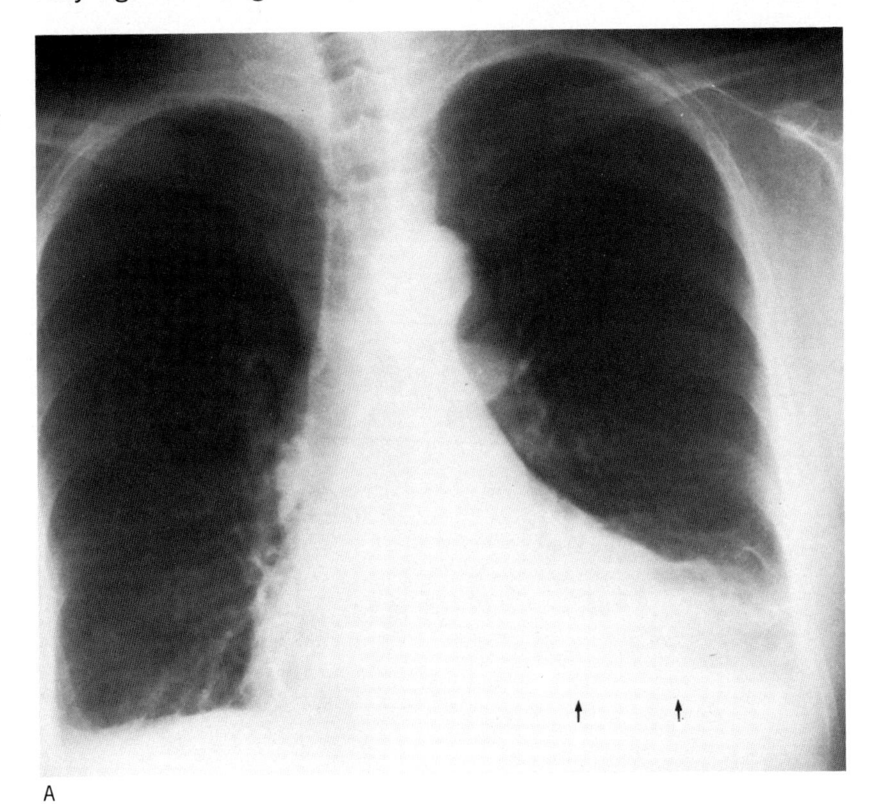

A

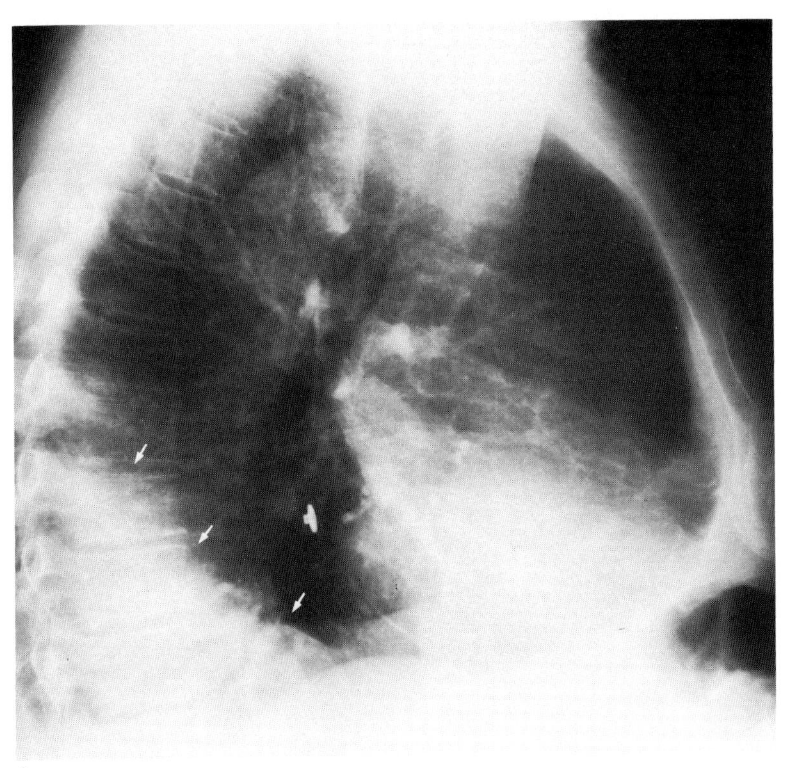

B

Fig. 2-20. *A.* PA view showing the density obliterating the shadow of the left hemidiaphragm (arrows). *B.* Lateral view showing the posterior density (arrows).

Fig. 2-21. *A.* The normal appearance of moderate right upper lobe collapse versus collapse with a tumor in the hilum producing an S sign of Golden (~). *B.* Arrows point to the superiorly displaced minor fissure in right upper lobe atelectasis. Contrast single curve, a lateral concavity appearance, of fissure to *C. C.* The concave lateral portion (white arrows) of the minor fissure and the convex portion with the mass in the S sign of Golden (black arrows).

the hilum appears small, because the lower lobe artery is silhouetted out and is displaced inferiorly.

The presence of a tumor mass in the hilum alters the appearance of collapse. The *S sign* of Golden [18] is seen whenever a tumor in the hilum produces a convex bulge on the collapsed upper lobe (Fig. 2-21). In the right middle lobe, the presence of a tumor changes the direction of the triangular wedge so that the apex points anteriorly away from the hilum on lateral view (Fig. 2-22).

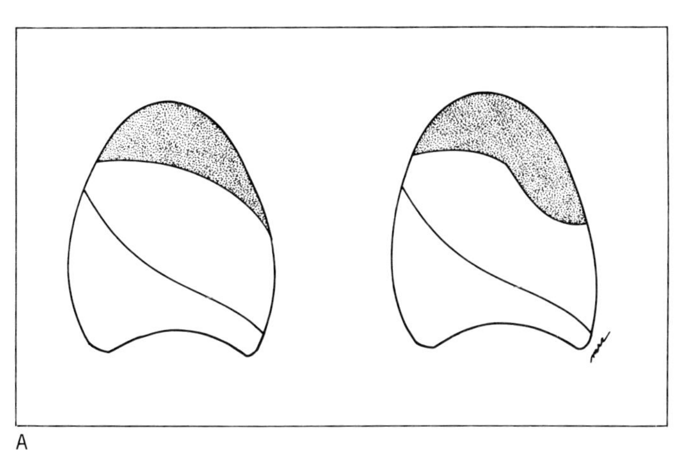

A

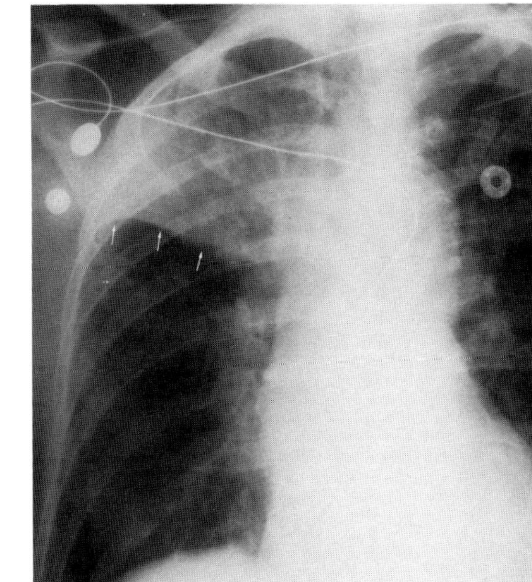

B

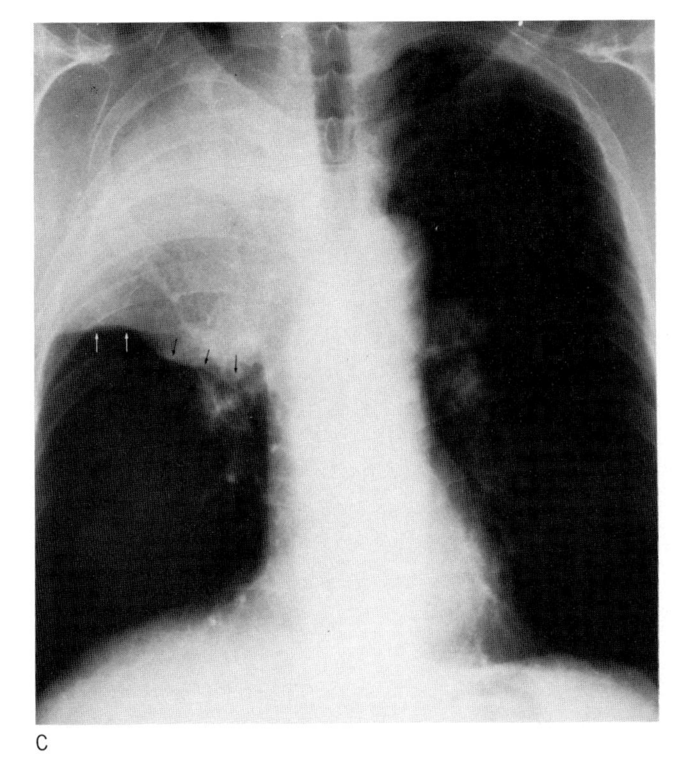

C

Subsegmental atelectases commonly occur. They vary from platelike (linear or discoid) atelectases [19], to patches of subsegmental density that cannot be differentiated from a pneumonic process, to a round atelectasis that can mimic a neoplasm. Round atelectasis [1,20] is seen following a pleural effusion. The floating end of a lung tilts on a cleft, and, as adhesions form, the lung is fixed in position despite resolution of the effusion. Round atelectasis is recognized by a characteristic curving of vessels and bronchi from its anterior inferior margin.

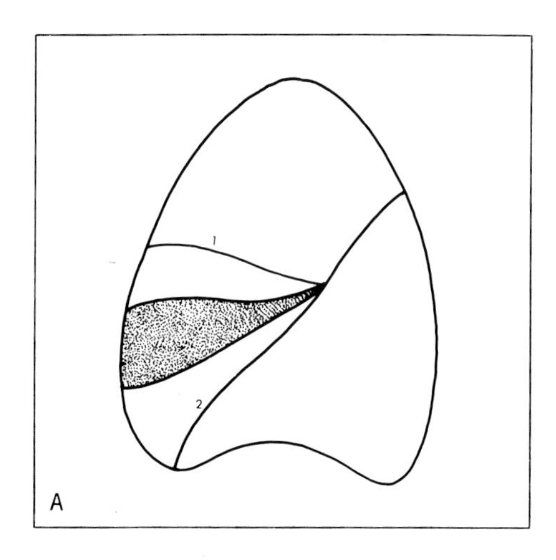

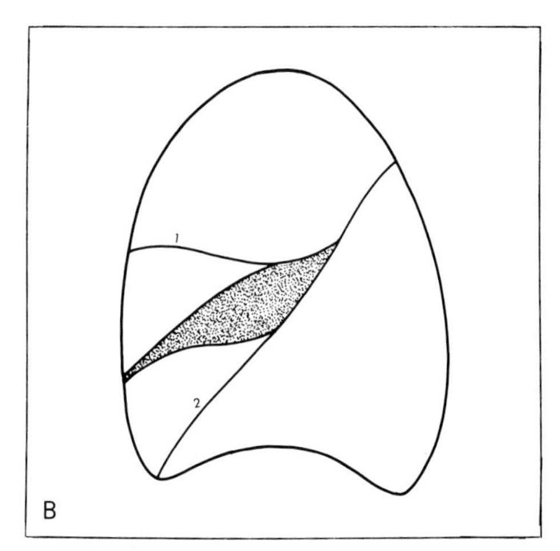

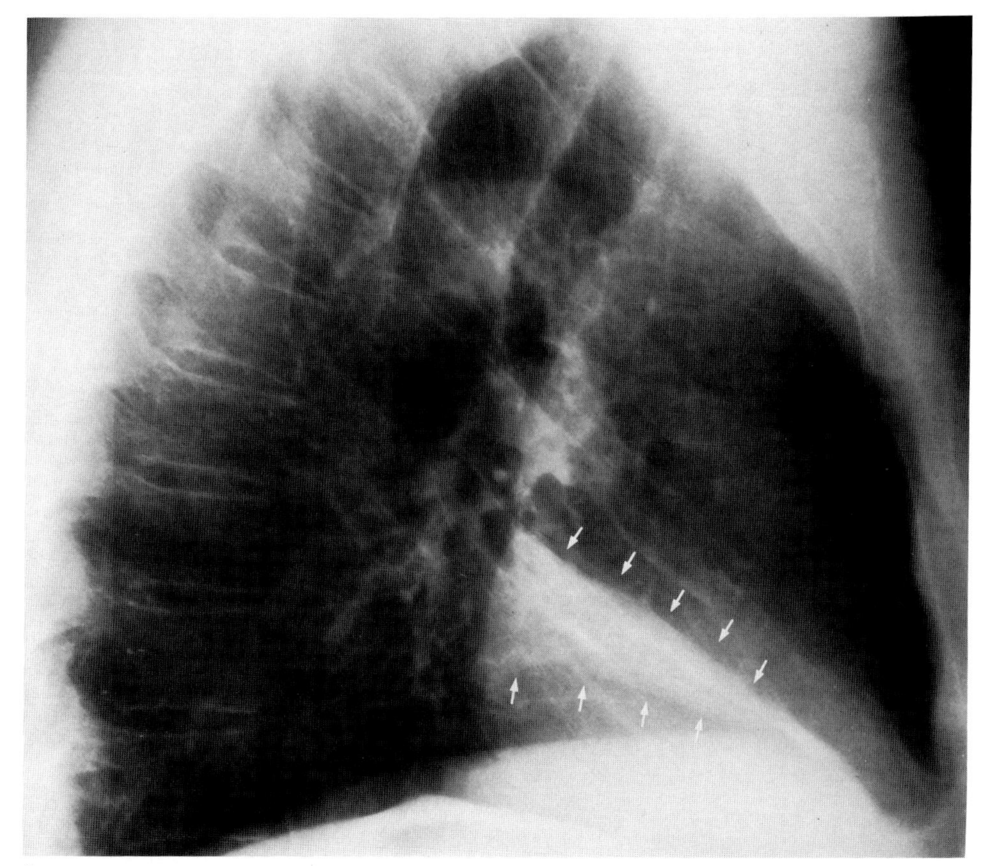

Fig. 2-22. *A.* Typical appearance of right middle lobe collapse. Line 1 represents the original position of the minor fissure; line 2 represents the original position of the major fissure. *B.* Appearance of right middle lobe collapse with a tumor in the hilum—the S sign of Golden equivalent for the middle lobe, a reversal of the position of apex and base of the triangular density produced by the collapsed lung. Wider portion is now in the hilum. *C.* Reverse wedge outlined by arrows in right middle lobe collapse with tumor in the hilum, the middle lobe equivalent of the sign of Golden.

PULMONARY NODULES AND MASSES

Pulmonary nodules and masses represent other densities seen on chest films. A nodule is a rounded, oval, or elongated density with fairly well-defined or well-circumscribed borders. Nodules may be benign (granulomas) or malignant (primary neoplasms or metastatic disease). Benign and metastatic nodules show smooth borders; primary malignant lesions tend to have ill-defined borders. Lobulation [21] (Fig. 2-23), umbilication [22,23,24], and the presence of a surrounding area of increased radiolucency from paracicatricial emphysema called *corona radiata* [25] (Fig. 2-24) all suggest malignancy. Satellite lesions can occur in both, but are more common in benign lesions [26]. Calcification is a sign of benignity only if it is homogeneous, central (Fig. 2-25), lamellated, stippled, or "popcorn" in appearance (Fig. 2-26). Eccentric calcification may be seen in both benign and malignant lesions. Other appearances previously considered to suggest malignancy include the presence of a pleural tail [27] (a thin, untapered, linear density extending from the periphery to a nodule and

Fig. 2-23. *A*. Lobulated mass in the left lower lobe. *B*. Lateral view of the patient in *A* showing the lobulated mass in the position segment of the left lower lobe.

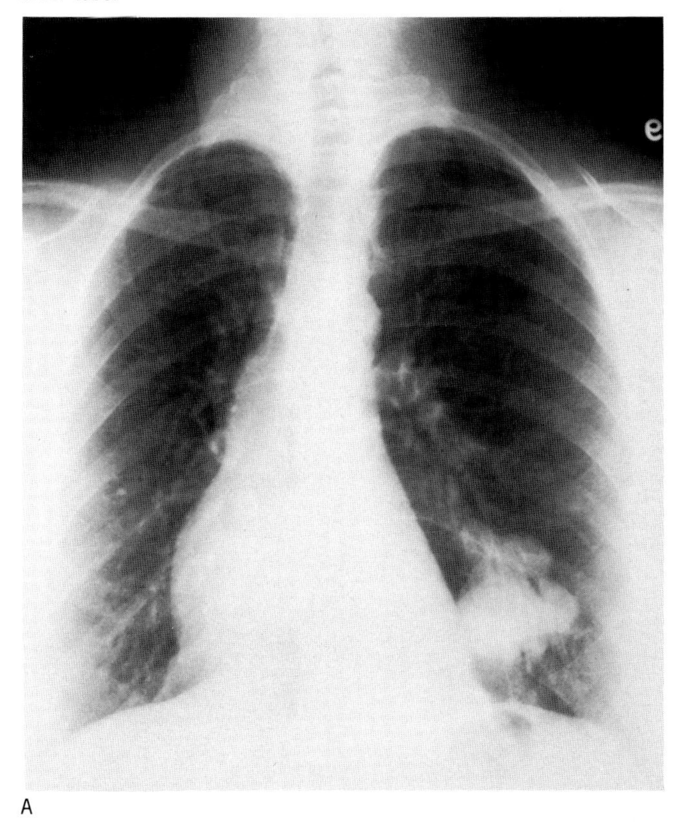

A

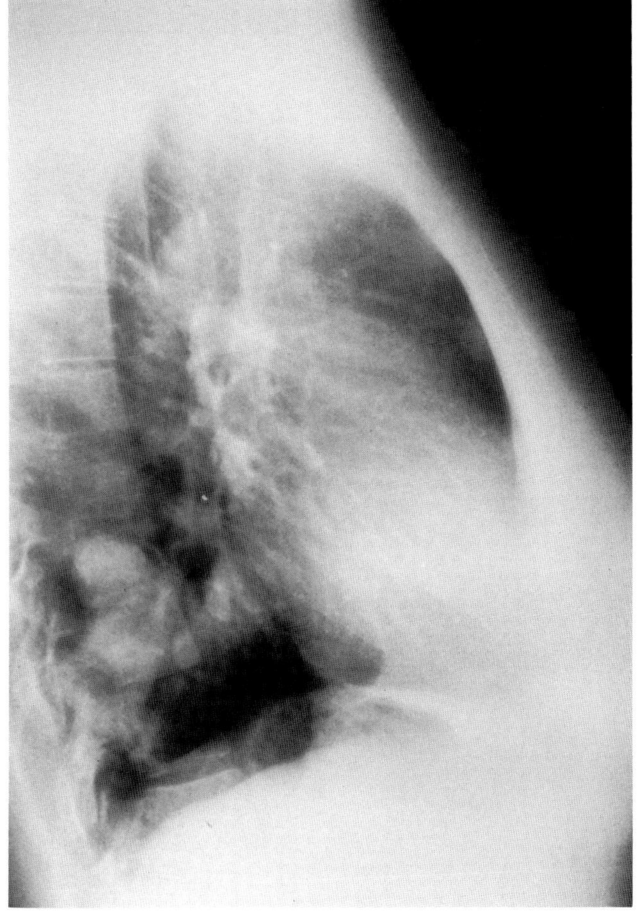

B

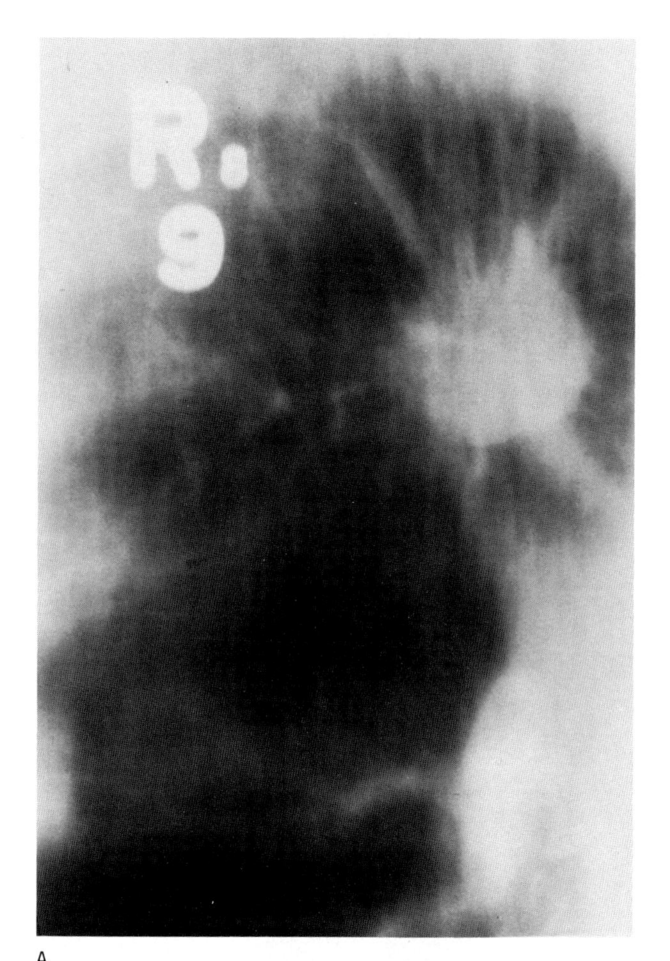

A

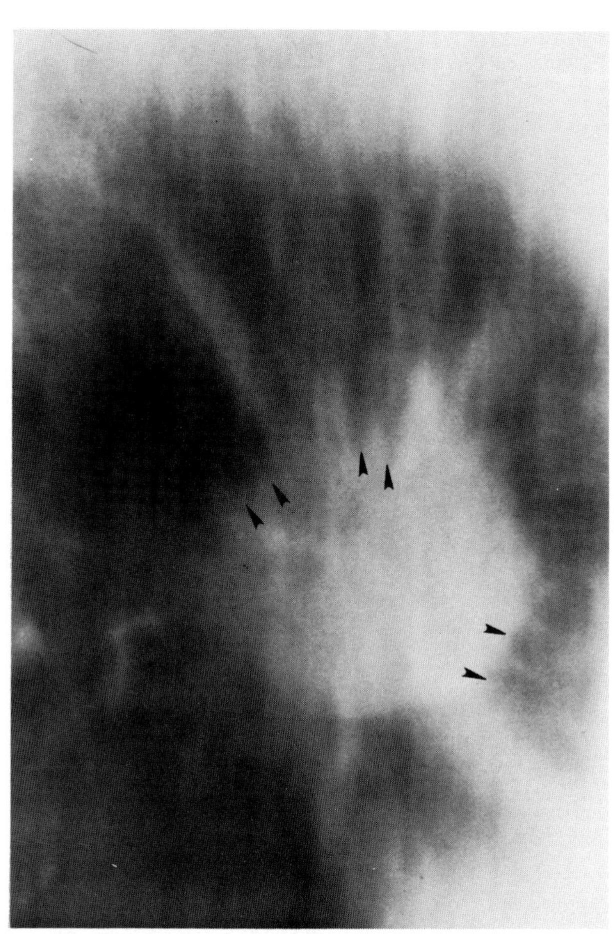

B

Fig. 2-24. *A.* Right upper lobe mass showing paracicatricial emphysema—corona radiata. *B.* Close-up of the paracicatricial area of emphysema (arrowheads) around the lesion. (Courtesy of Dr. Jerry Balikian.)

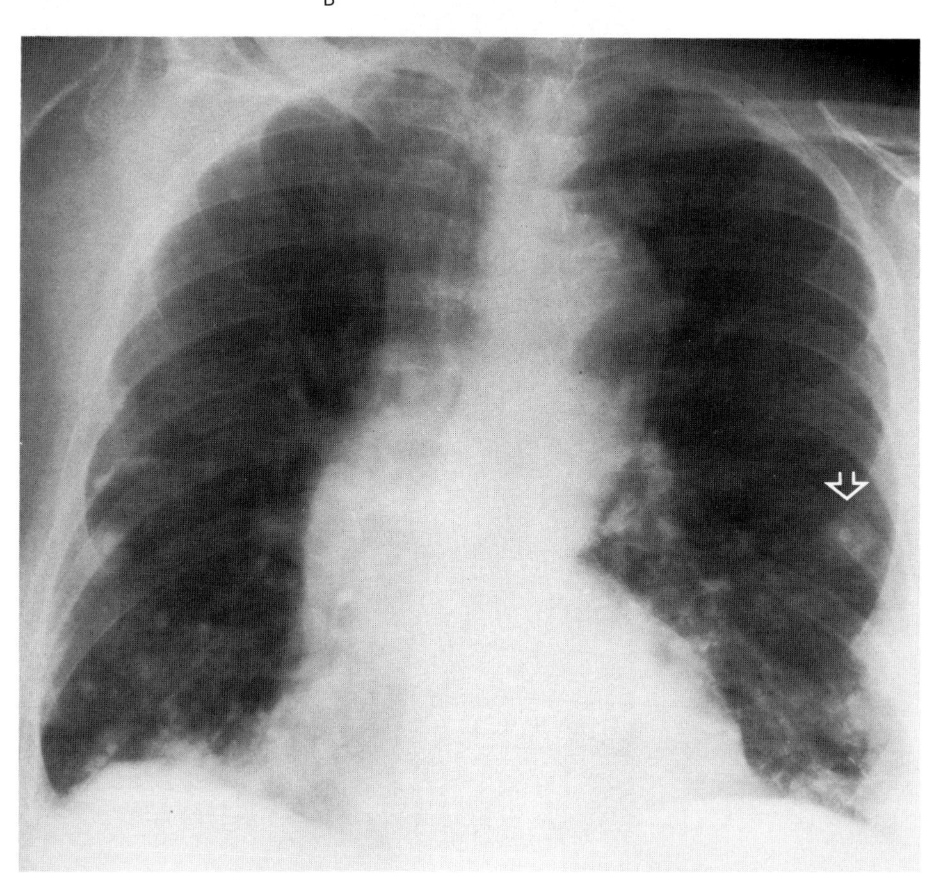

Fig. 2-25. Multiple granulomas. Open arrow points to a nodule with central calcification.

Fig. 2-26. *A*. Popcorn calcification in a hematoma (arrow). *B*. Lateral view of the patient in *A*.

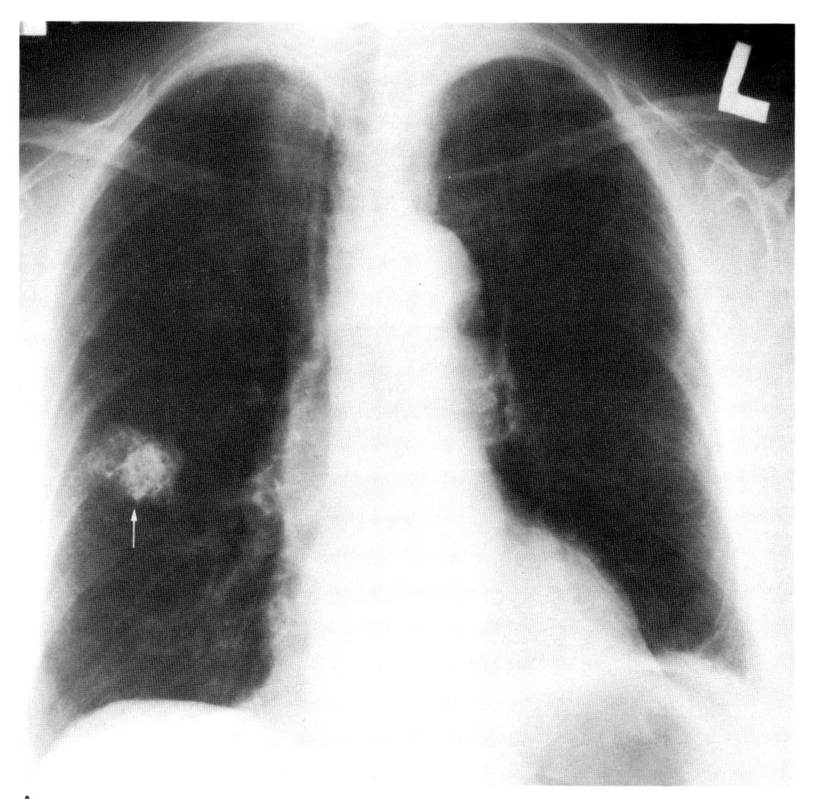

A

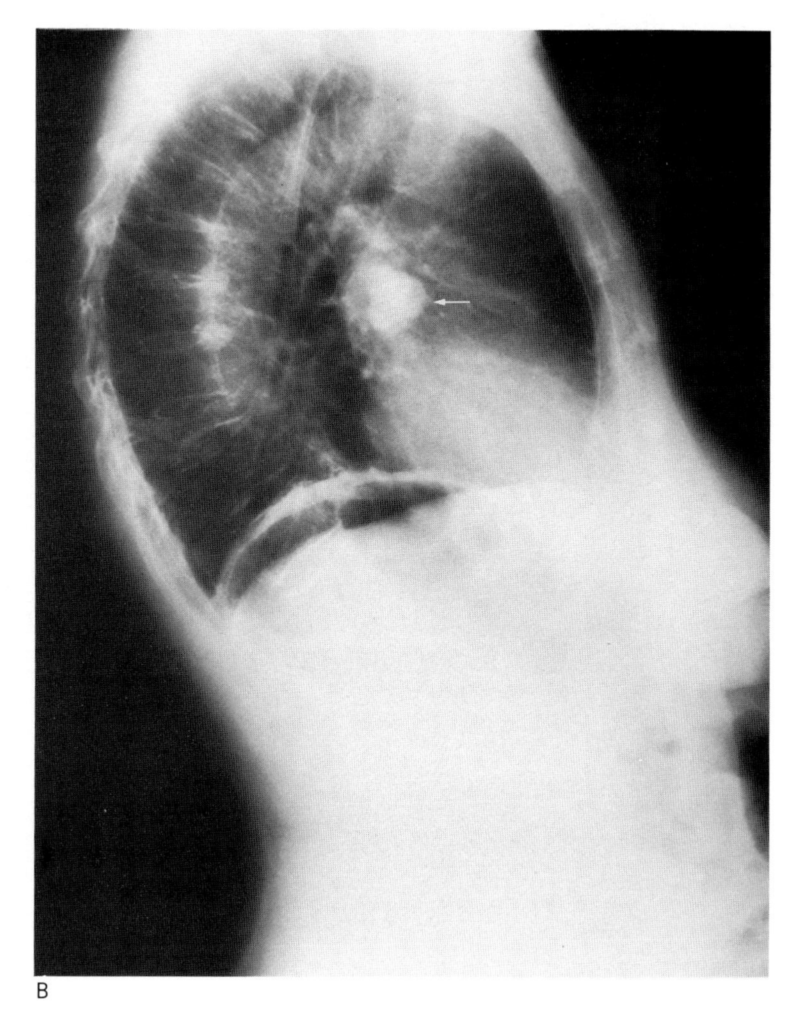

B

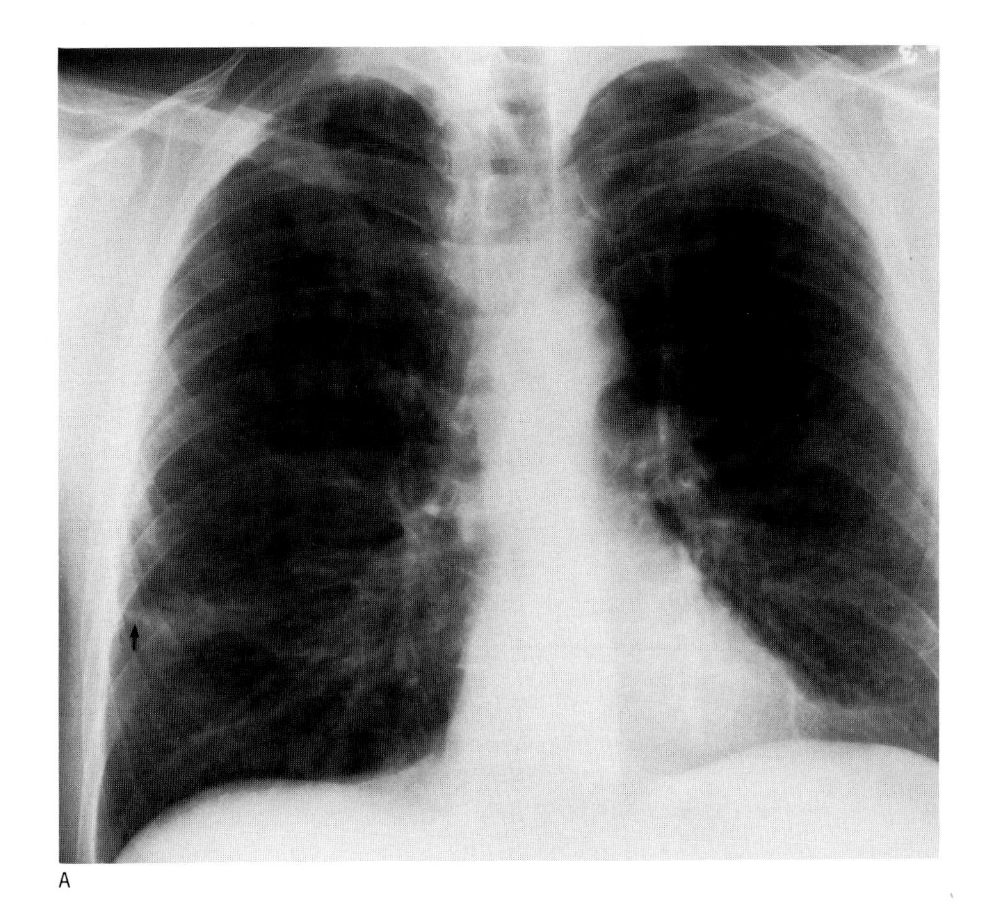

A

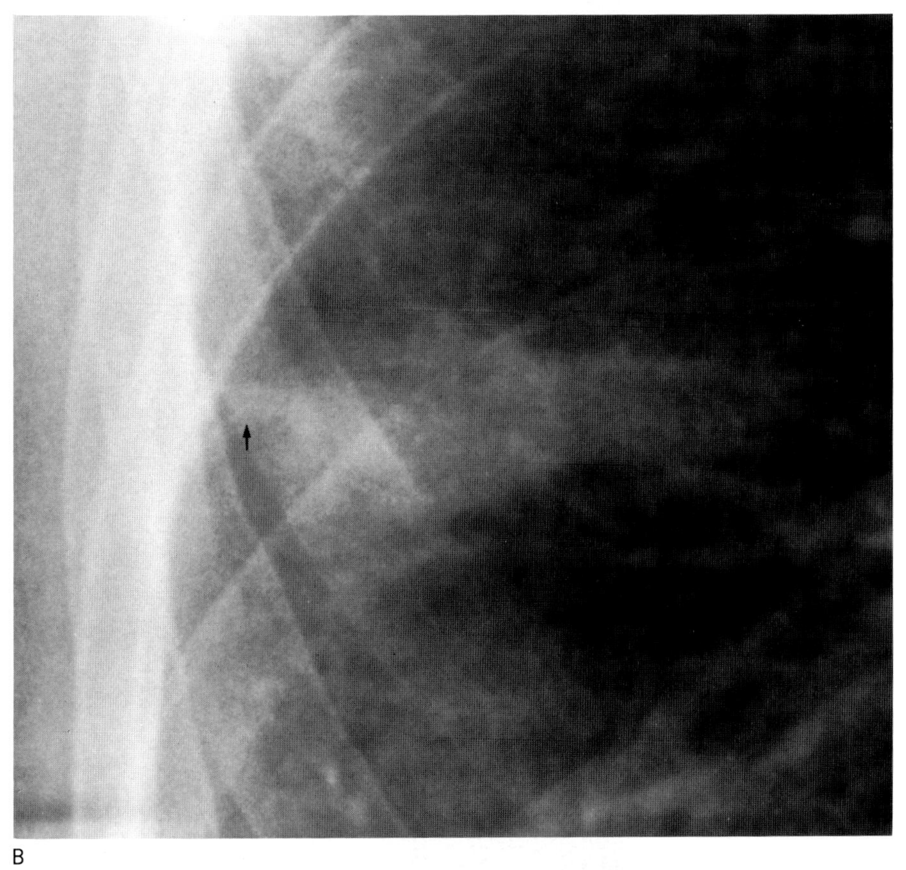

B

Fig. 2-27. *A.* Irregular pulmonary nodule with a pleural tail (arrow). *B.* A close-up of the nodule. Arrow points to the pleural tail.

ending as a triangular density at the pleural surface, representing a fibrotic change with indrawing of pleura), several pleural tails, or a thicker shadow of indrawn pleura extending into the mass (Fig. 2-27). However, these appearances are now known to be present with equal frequency in benign lesions [28,29]. Peripheral malignant neoplasms tend to be located nearer the hilum as they grow [30].

When only part of a nodule is well defined and the rest of the nodule is poorly defined or nonvisualized, it is most likely a skin nodule and not intraparenchymal in location. This appearance is especially useful in differentiating between nipple shadows and nodules in the vicinity of the breast. Nipples are rounded densities that usually show a well-defined lateral edge and an ill-defined inner margin (Fig. 2-28).

Tiny (1–2 mm) nodules similar in size to millet seeds are called miliary nodules. Typically caused by miliary tuberculosis (Fig. 2-29), other diseases such as pneumoconiosis (silicosis, coal-worker's pneumoconiosis), sarcoidosis, pulmonary hemosiderosis, Goodpasture's syndrome, and alveolar microlithiasis also produce them. Microlithiasis can be differentiated from the others by the increased density of the nodules and the presence of a negative, or black, pleural line (Fig. 2-30) instead of a normal white pleural line [7].

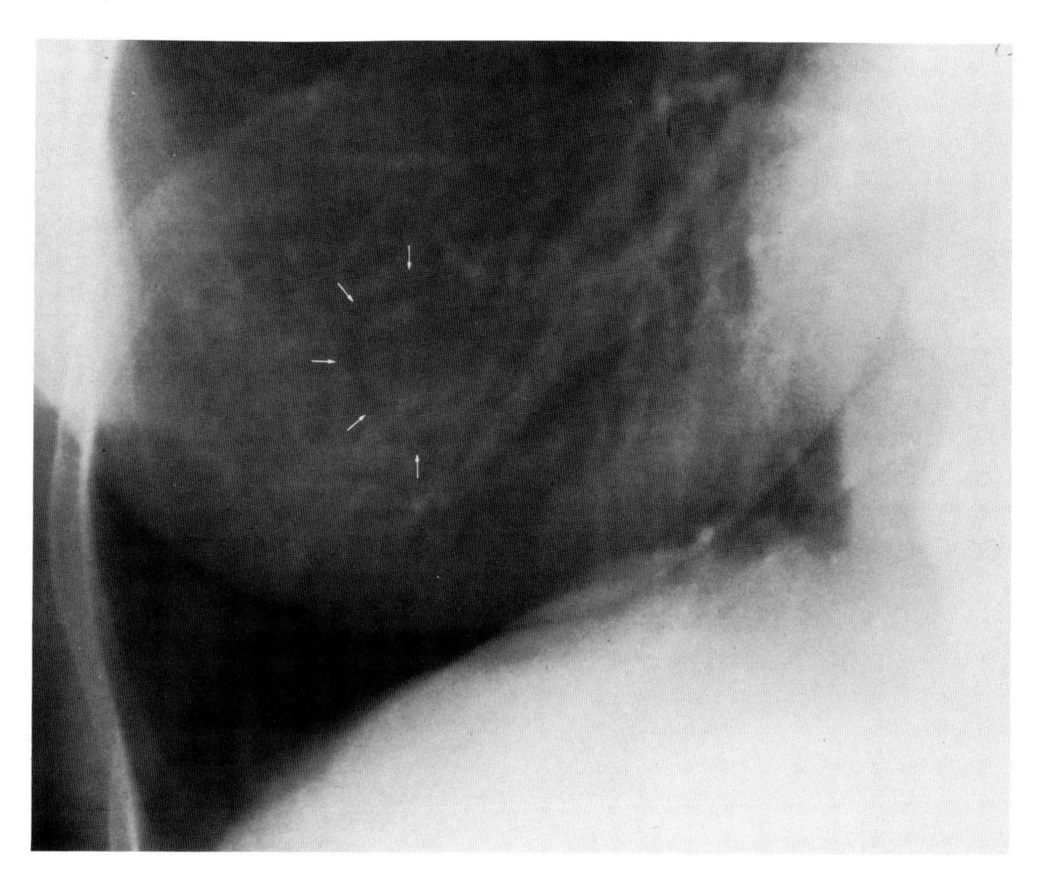

Fig. 2-28. Close-up of nipple shadow with a lucent rim (air around nipple outlined by arrows) in its lateral border and an ill-defined medial border.

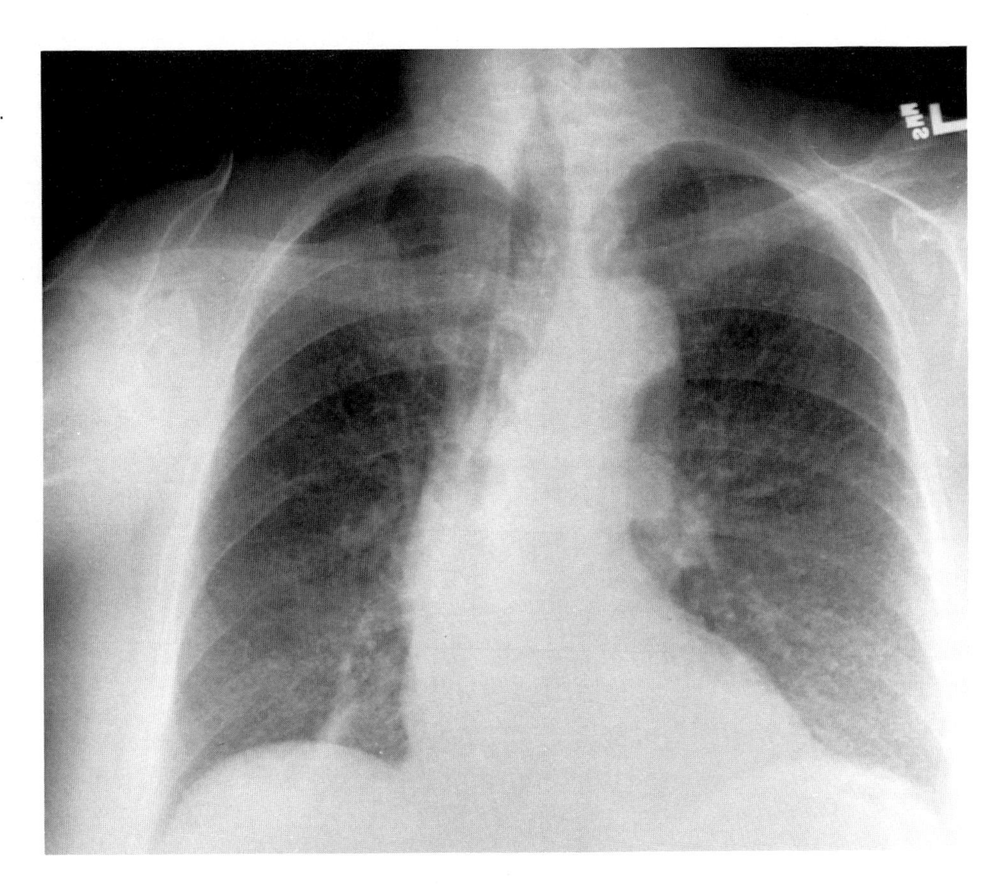

Fig. 2-29. Fine millet seed–sized pulmonary nodules in a patient with miliary tuberculosis.

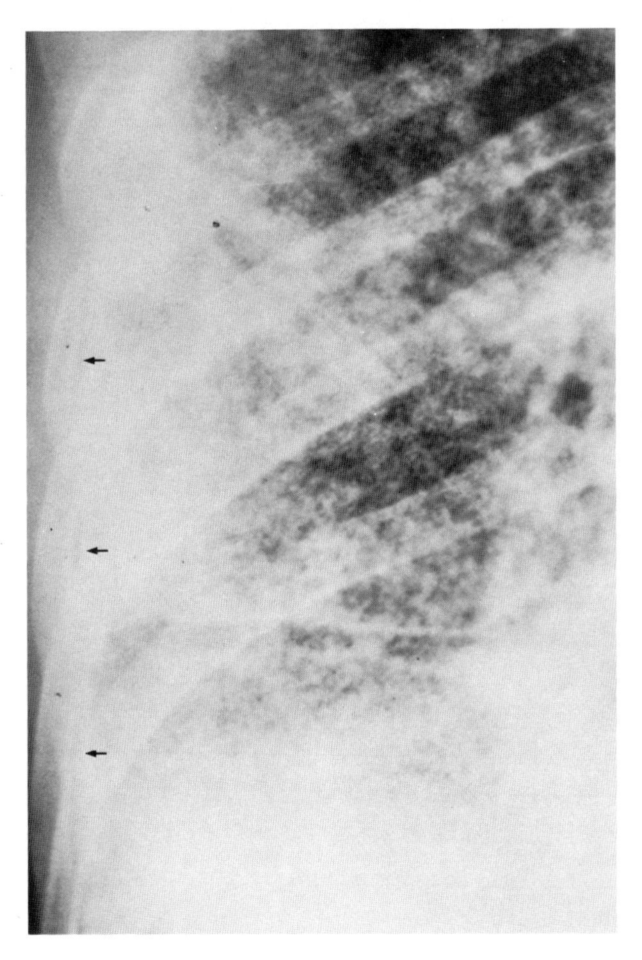

Fig. 2-30. Arrows point to the black pleural line in alveolar microlithiasis.

Fig. 2-31. *A*. Two- to five-mm pulmonary nodules in a patient with sarcoid. *B*. Close-up of nodules.

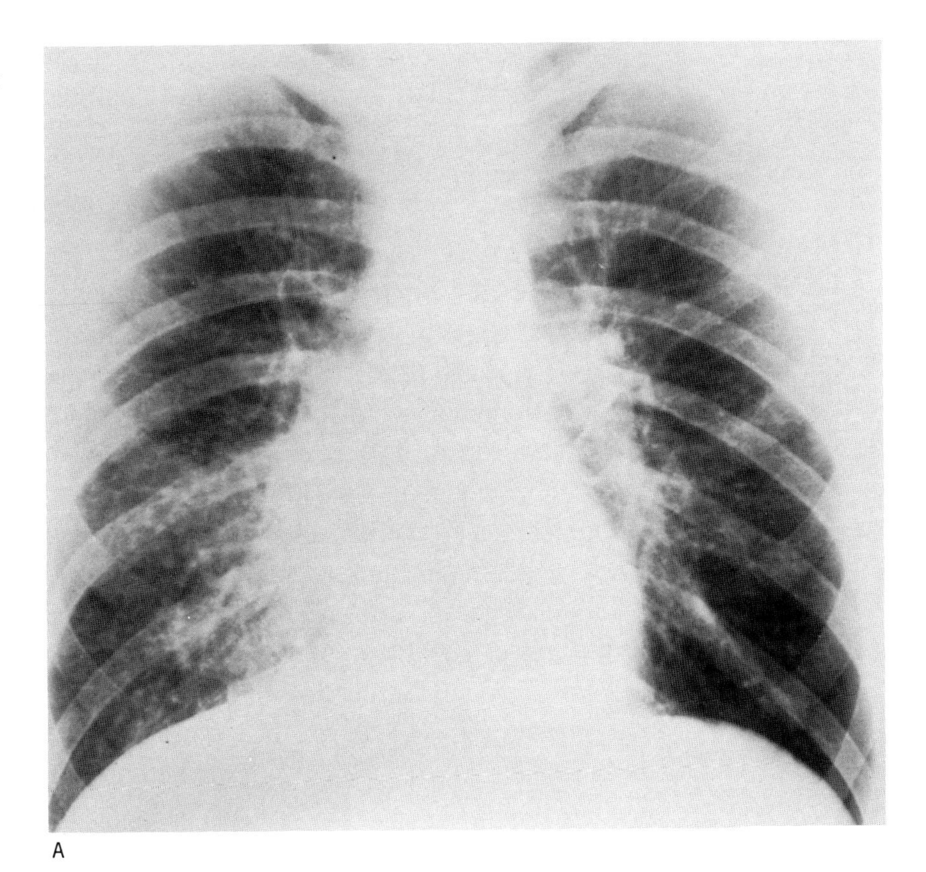

A

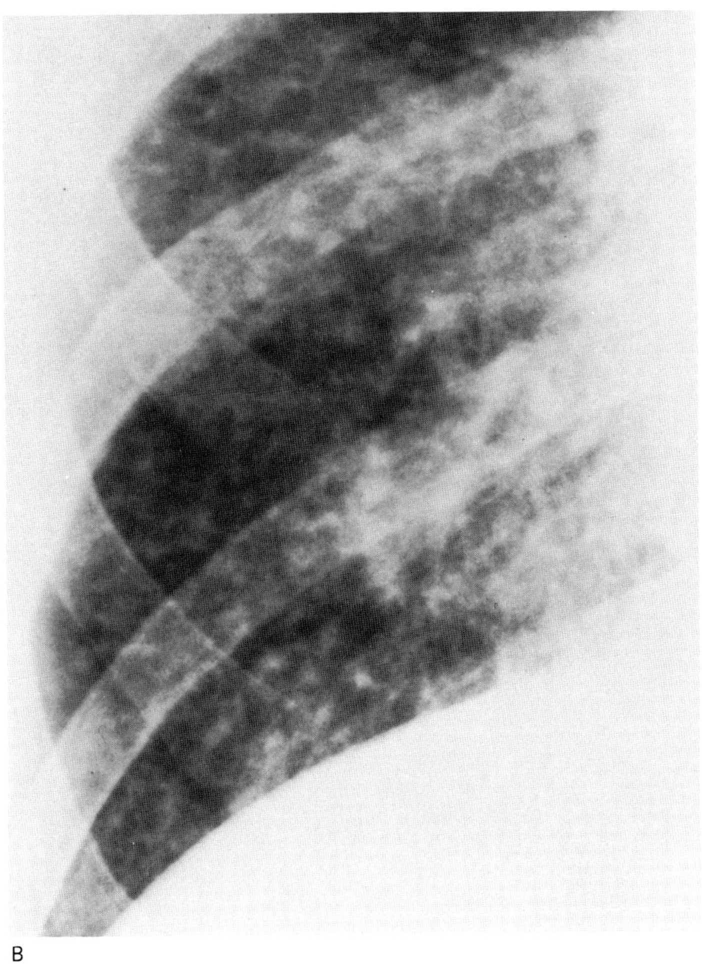

B

Small nodules (2–5 mm) are also seen in some of the above diseases, such as pneumoconiosis and sarcoidosis (Fig. 2-31), but the differential diagnosis is greater and includes varicella pneumonia (Fig. 2-32), eosinophilic granuloma, alveolar cell carcinoma (Fig. 2-33), alveolar proteinosis, metastasis (from breast, thyroid, renal cell carcinoma [Fig. 2-34], and melanoma), congestive failure, and septic emboli.

Larger nodules (> 5 mm) are often produced by metastatic disease (e.g., colon carcinoma, Wilms' tumor, osteogenic sarcoma), rheumatoid nodules, Wegener's granulomatosis (Fig. 2-35), tuberculosis, histoplasmosis, primary neoplasms both benign and malignant, and fungal lesions.

Extrapleural lesions, except when seen *en face,* usually show a sharp inner margin and a poorly defined outer margin. In addition, the angle formed by the mass or nodule with the chest wall aids in determining whether it is pulmonary or pleural/extrapleural in location (Fig. 2-36). Lung lesions show acute angles (Fig. 2-37A); pleural and extrapleural lesions, obtuse angles (Fig. 2-37B). Except when seen *en face*, a pleural/extrapleural lesion also shows poor definition of a portion of its border. Unlike the nipple, pleural lesions show a sharp, convex inner border and a poorly defined lateral border. The edge of pleural lesions is tapered instead of rounded.

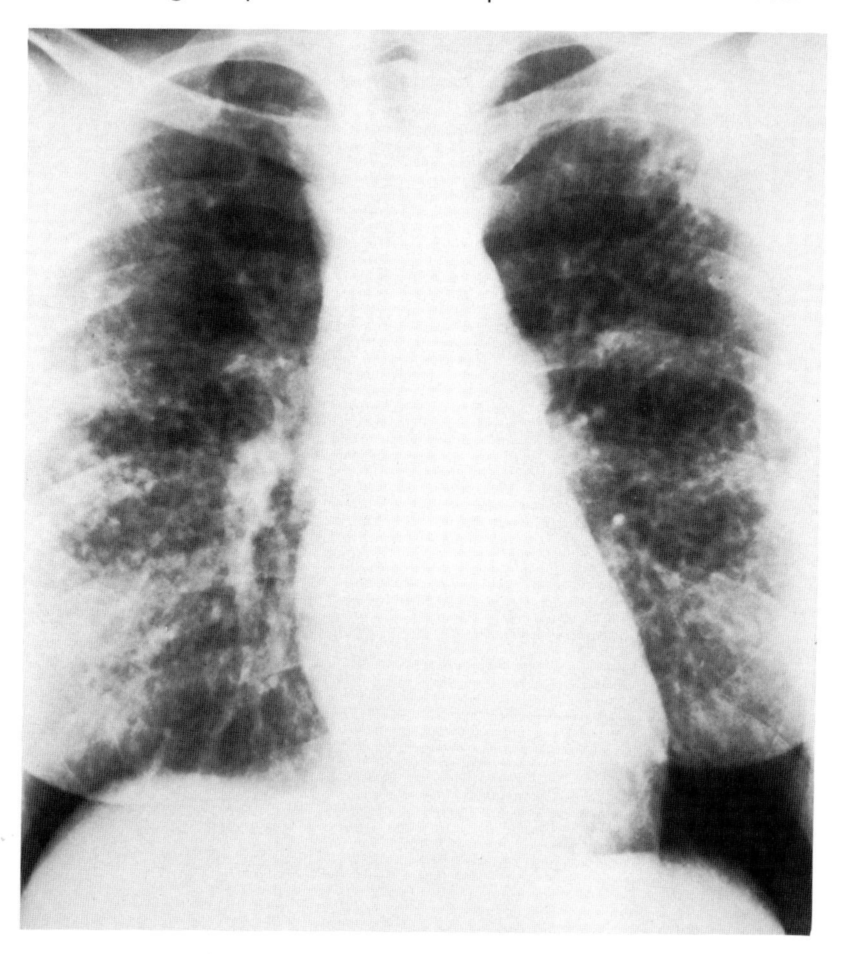

Fig. 2-32. Pulmonary nodules from varicella pneumonia.

Fig. 2-33. *A.* PA view of multiple pulmonary nodules from bronchoalveolar cell carcinoma. *B.* Lateral view of patient in *A.*

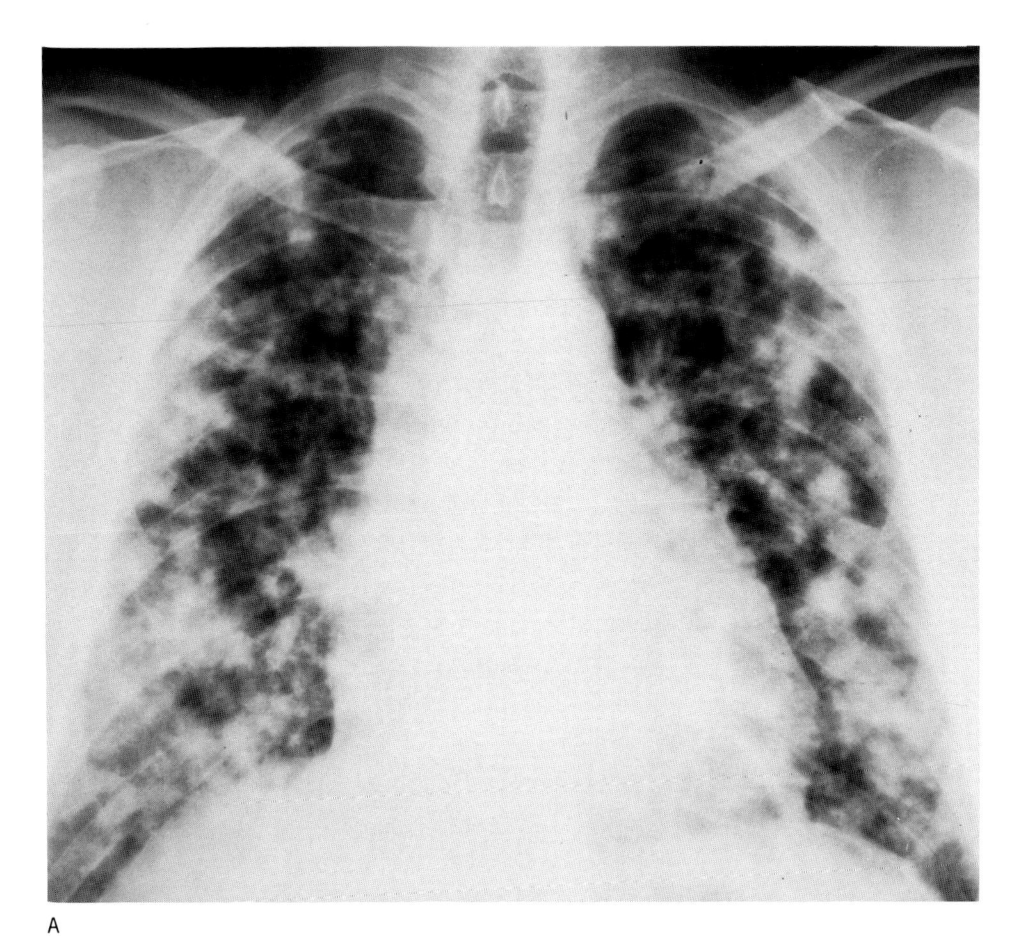

A

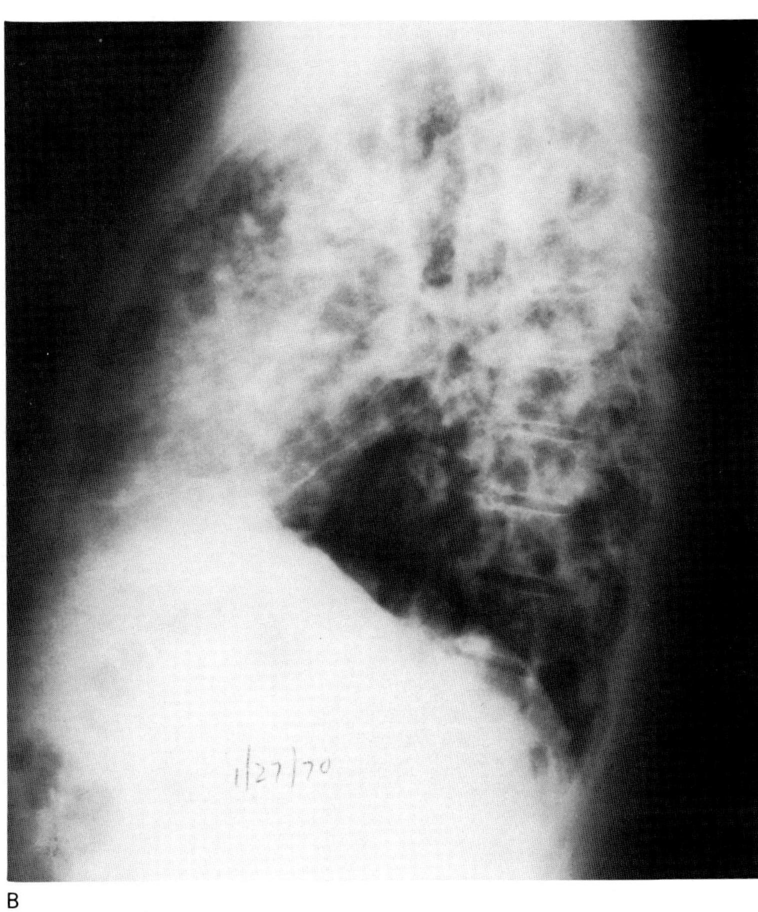

B

Fig. 2-34. Multiple 2- to 5-mm pulmonary nodules.

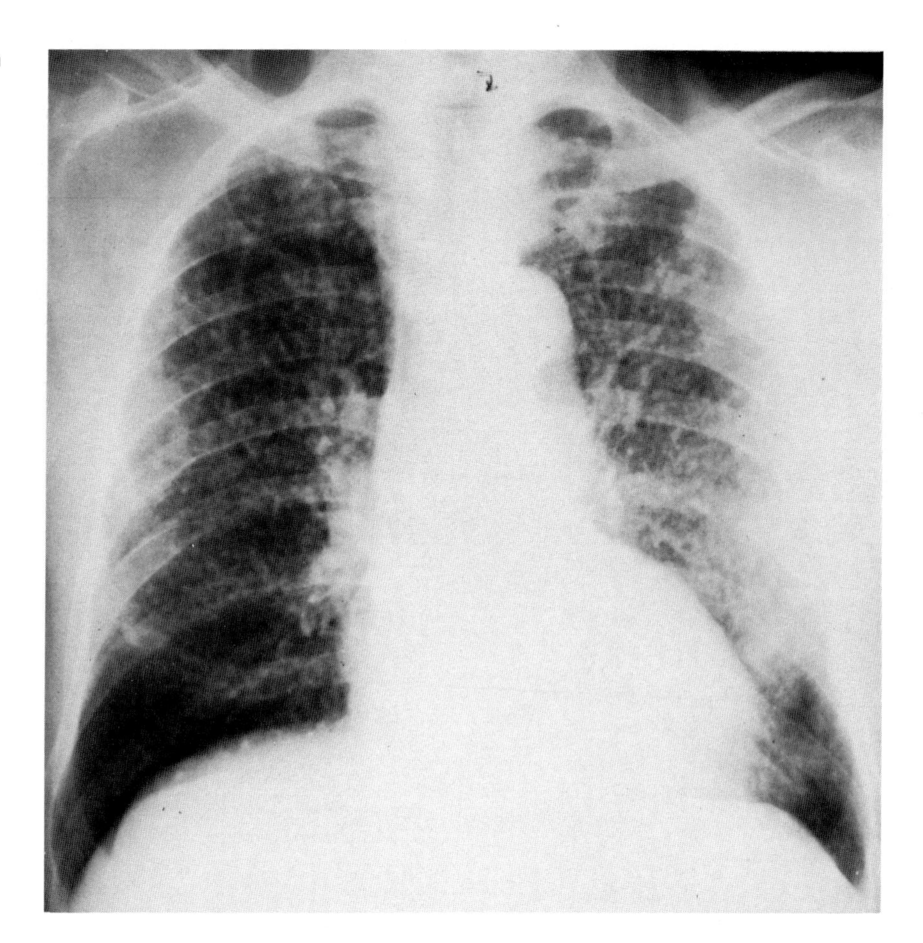

Fig. 2-35. *A*. PA view of a patient with large pulmonary nodules from Wegener's granulomatosis. *B*. Lateral view of patient in *A*. *C*. Multiple variably sized pulmonary nodules with the largest lesion in the left lower lobe in a patient with colon carcinoma.

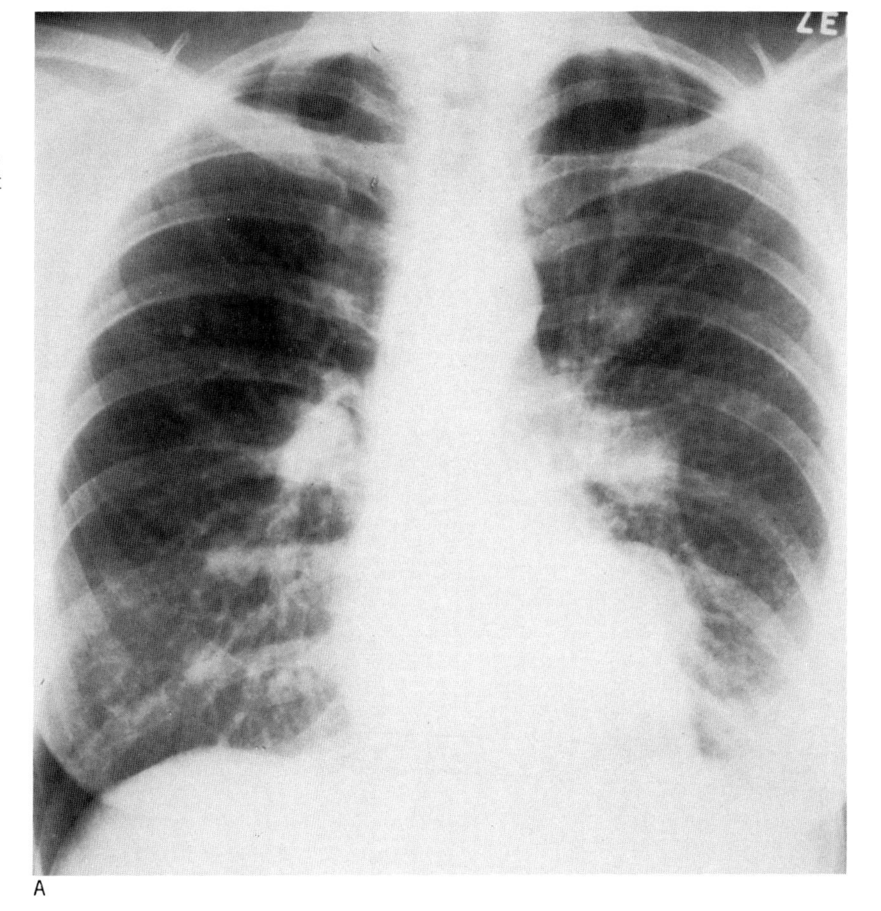

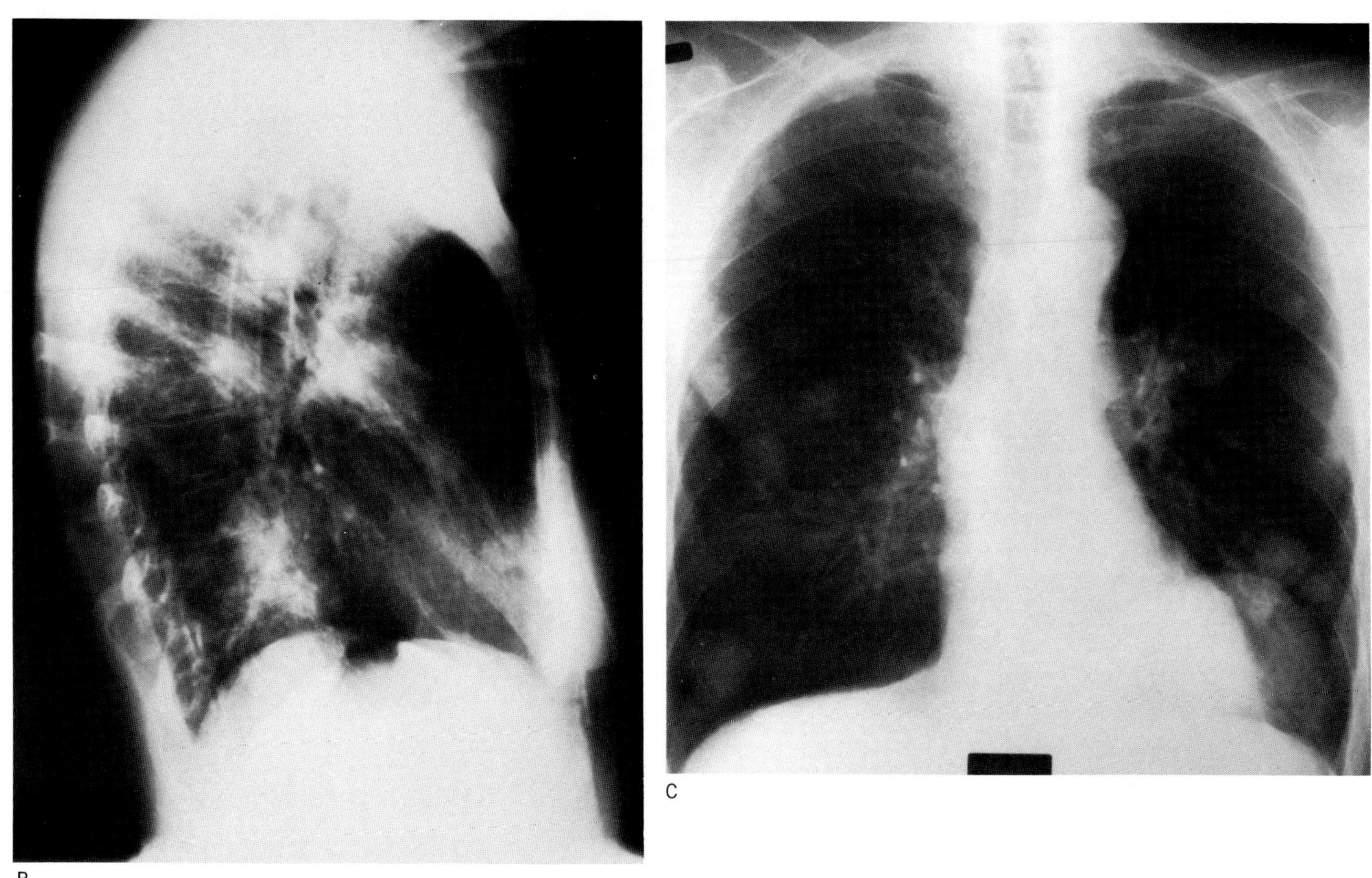

Fig. 2-35. (continued)

Fig. 2-36. Acute angle (arrow 1) produced by a parenchymal density abutting the pleura. Obtuse angle (arrow 2) produced by a pleural/extrapleural density.

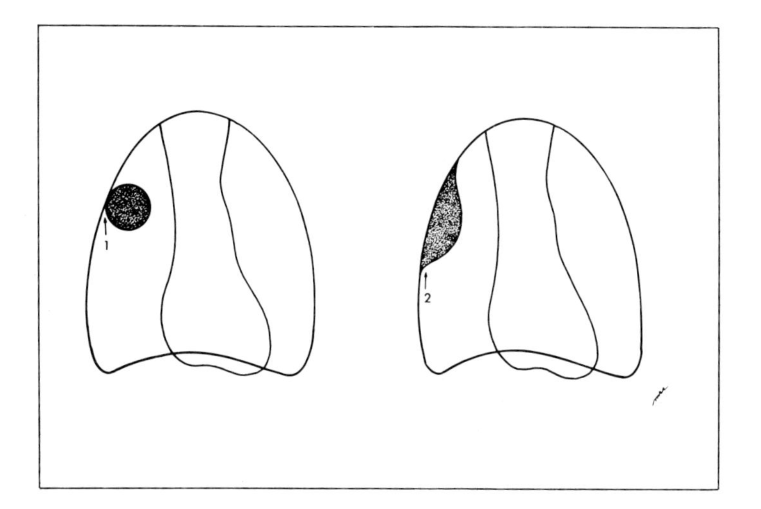

Fig. 2-37. *A.* Large rounded right lower lobe density (arrow). Note acute angle (arrowhead) formed by the lesion. *B.* A pleural density in the right upper lobe (arrow). *C.* Obtuse angle produced by the pleural lesion (arrows). *D.* The pleural lesion with a density measurement of 171 HU, a pleural lipoma.

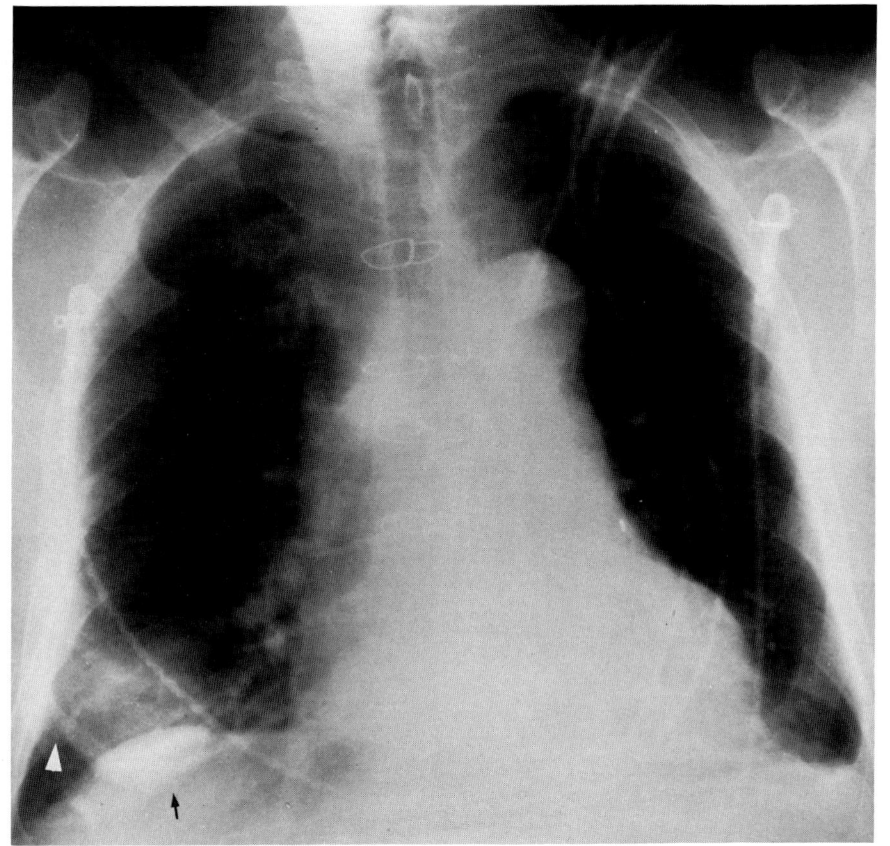

A

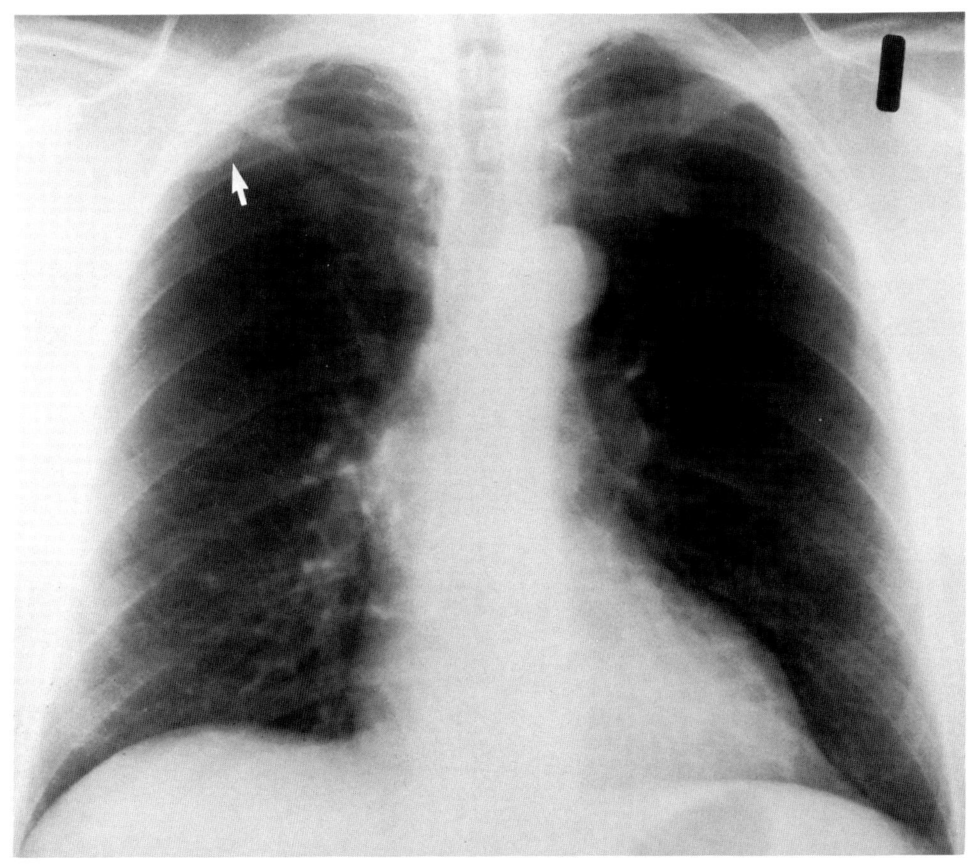

B

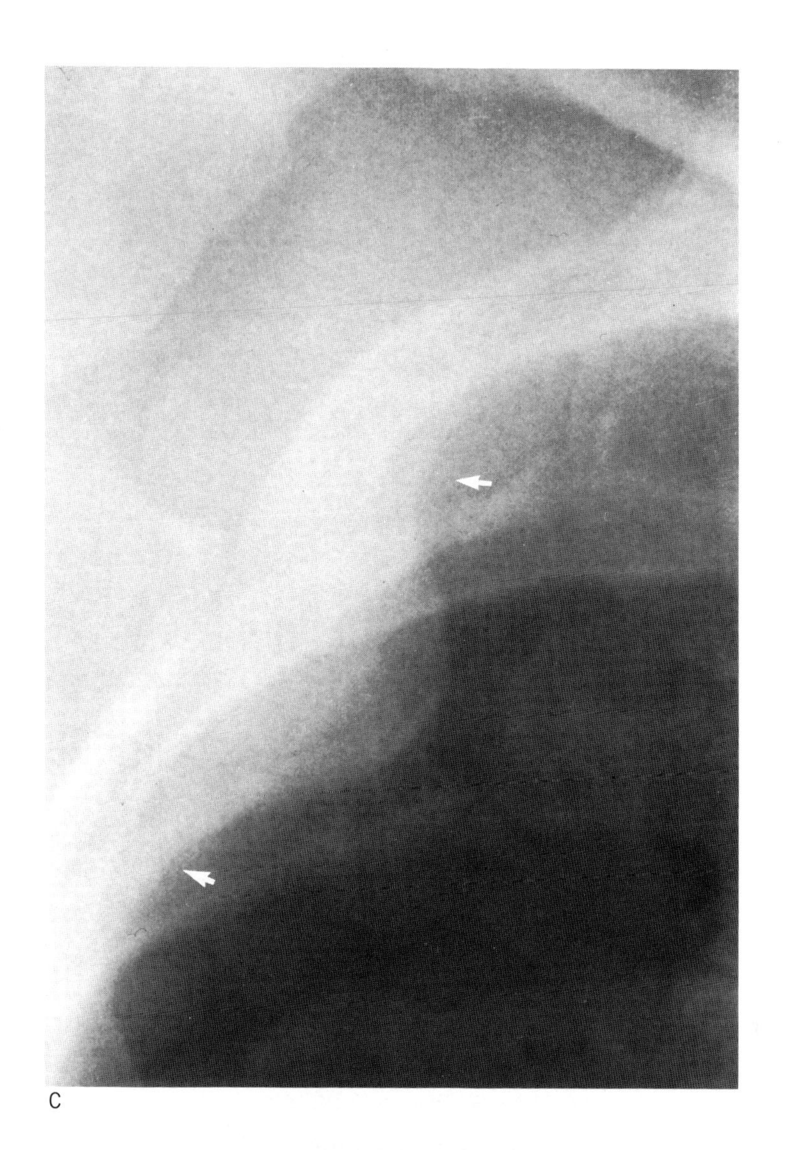

C

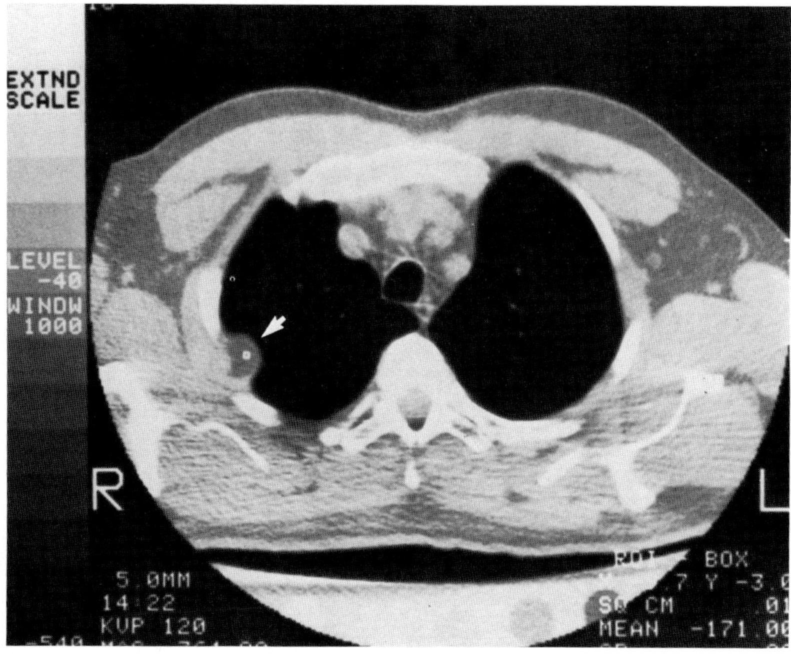

D

This combination of a sharp, convex outline and tapering margins is called the extrapleural sign of Felson [9]. Proper obliquity through fluoroscopy or lateral projection may be necessary to see the tapered edge of the lesion. In addition, movement of the lesion is synchronous with the overlying rib and independent of the lung.

Although the angle formed by the lesion with the chest wall determines its location (pleural/extrapleural lesions form obtuse angles and parenchymal lesions form acute angles), the angles formed are not 100 percent accurate. Pedunculated pleural lesions may indent the pleura and produce acute angles. A parenchymal pulmonary lesion that has extended to the pleura may produce the obtuse angle seen in pleural lesions. Other secondary signs such as an air bronchogram, linear strands arising from the lesion, or an irregular, shaggy border suggest pleura-based parenchymal lesions such as pulmonary infarcts, Pancoast's tumors, lymphomas, pseudolymphomas, metastases, fungal lung lesions, and granulomatous lesions.

Tissues in the pleural/extrapleural region may give rise to both benign and malignant neoplasms. Table 2-2 lists the possible pleural/extrapleural lesions.

What criteria should one use in the management of pulmonary nodules?

1. Presence of the nodule in a film taken 2 years or more previously without a change in size suggests benignity.
2. In young patients (under 35 years) without known primary malignancies, a pulmonary nodule is considered benign, and a follow-up film should be obtained in 6 to 12 weeks, and another in 3 months. Subsequently, a film should be obtained at 6-month intervals for the next 18 months.
3. In patients over 35, a nodule should be considered malignant until proven otherwise.
4. Eccentric calcification in a patient at any age should be considered malignant (scar carcinoma) until proven otherwise.

If a nodule that is being followed closely shows a change in size, one should evaluate the doubling time of the lesion (the time it takes the lesion to double in volume, not diameter). If the radius of the lesion is multiplied by 1.25, the product is the volume of the lesion after it has doubled. Nodules with a doubling time of 7 days or less, or of 465 days or more, are considered benign [31]. Most lesions with a doubling time of less than 37 days are also benign. Malig-

TABLE 2-2. Pleural/Extrapleural Lesions

	Neoplasms	
Tissue source	Benign	Malignant
Mesothelium	Mesothelial cyst	Mesothelioma
Muscle	Rhabdomyoma	Rhabdomyosarcoma
Fibrous tissue	Fibromas Desmoid tumors Organizing hematomas Empyema Loculated effusion	Fibrosarcoma
Fat	Lipoma	Liposarcoma
Intercostal nerve	Neurofibroma Schwannoma	Neurofibrosarcoma
Vessels	Hemangioma	Hemangiosarcoma
Bone	Osteochondroma Fibrous dysplasia Eosinophilic granuloma Granulomatous infections* Fracture/hematoma	Ewing's sarcoma Multiple myeloma Metastases

These lesions show rib destruction associated with the pleural/extrapleural masses, often giving a clue to the etiology of the pleural/extrapleural lesion.
*Actinomycosis, blastomycosis, aspergillosis, nocardiosis, tuberculosis.

nant lesions that double with 37 days are metastatic choriocarcinoma, testicular carcinoma, osteogenic sarcoma [32], and Burkitt's lymphoma.

Calcification in pulmonary nodules is found in both benign and malignant lesions. Benign calcifications are found in hamartomas, granulomas from tuberculosis, histoplasmosis, varicella infection, coccidioidomycosis, and pulmonary blastoma. Benign ectopic calcifications are found in nodules in patients with renal failure or on dialysis [33].

Of the malignant lesions, calcifications are seen in metastasis from mucinous adenocarcinoma of the gastrointestinal tract, bone metastasis from osteogenic sarcoma, chondrosarcoma, rarely giant-cell tumor, synovial sarcoma, thyroid carcinoma, ovarian papillary cystadenoma, and metastatic nodules that have bled or necrosed following radiotherapy or chemotherapy [34].

Cavitation of a pulmonary nodule or mass may occur. Among the more common causes are abscess, septic emboli, bronchogenic carcinoma (especially squamous cell), metastatic lesions (especially squamous cell carcinoma of the genitourinary tract [Fig. 2-38], head, and neck), hematoma, infarct, lymphoma, rheumatoid nodule, Wegener's granuloma, and mycobacterial, fungal (Fig. 2-39), and hydatid infections.

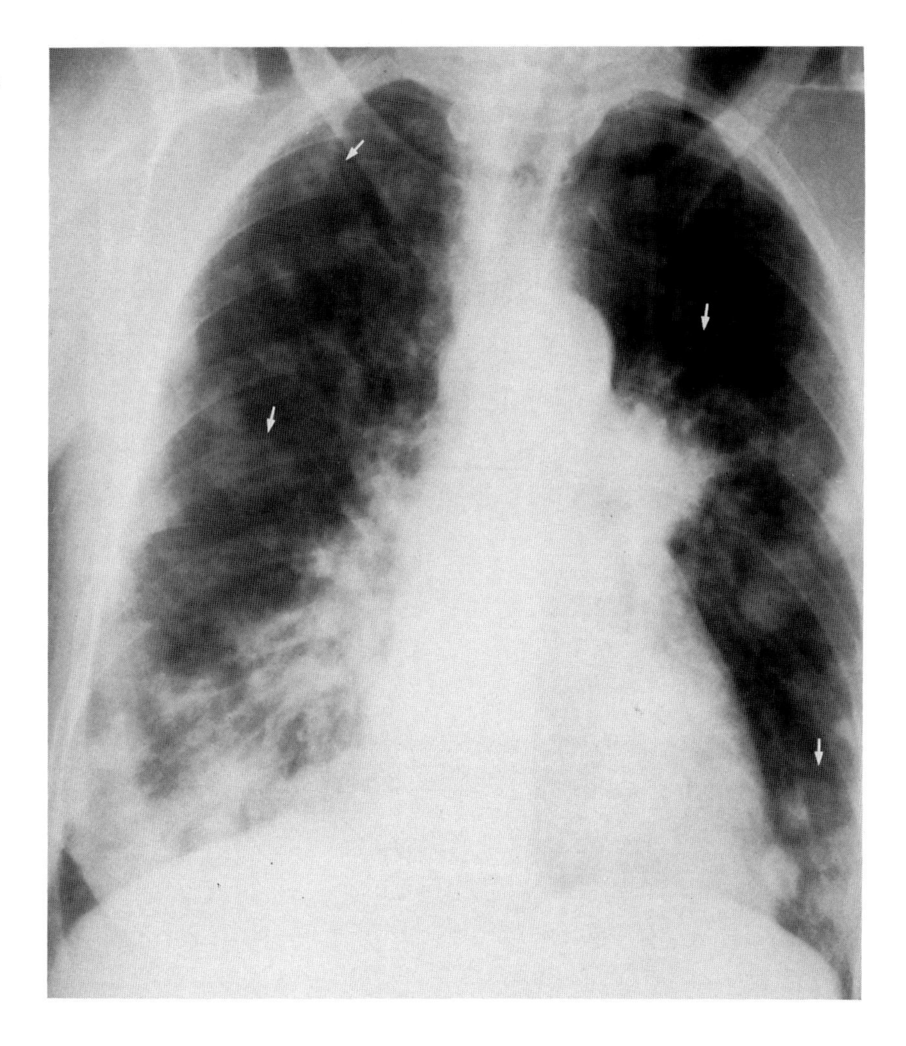

Fig. 2-38. Some pulmonary metastatic nodules with cavities (arrows) from carcinoma of the vulva.

Fig. 2-39. *A*. Focal infiltrates (arrows) in the right upper and left lower lobes. *B*. Arrows point to crescent-shaped lucencies (air) developing within the lesions, 6 days later.

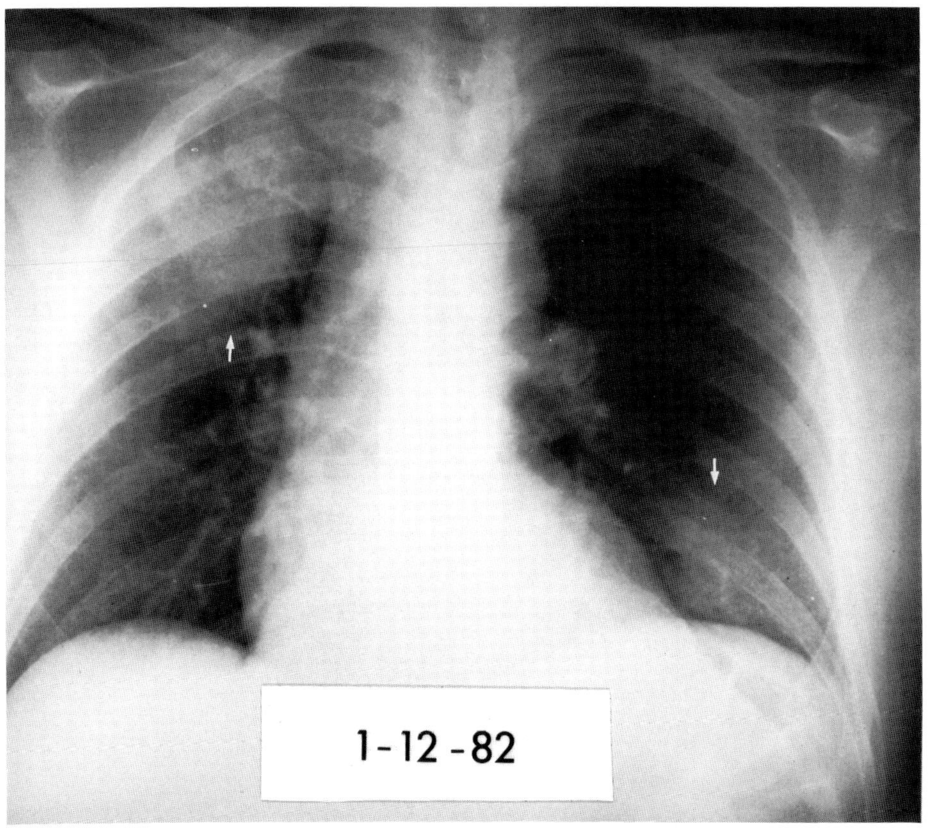

A

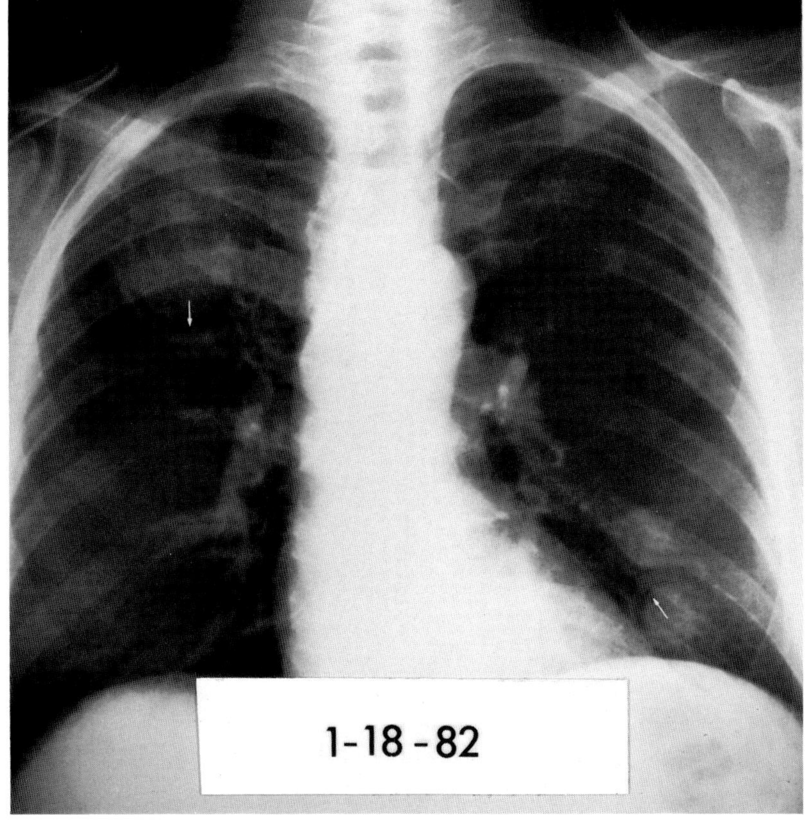

B

CAVITIES AND CYSTS

An air-filled or fluid-filled abnormal parenchymal space with definite walls represents either a cavity or a cyst. Cysts may be congenital or acquired; cavities are acquired and are the result of inflammatory or neoplastic processes. Cavities and cysts may be single or multiloculated, and round, oval, or elliptical.

It is important to analyze the walls and contents of cavities. Walls may be thin or thick, smooth or irregular. The contents may be pure air, air and fluid, or a mass of tissue. Dense, thick, and irregular walls (Fig. 2-40) are usually seen in the following conditions:

1. Neoplasms that have undergone cavitation. Primary bronchogenic squamous cell carcinoma (Fig. 2-41), papilloma in children, lymphoma, metastatic carcinoma, and sarcoma cavitate. Of the metastatic carcinomas, squamous cell carcinomas of the head and neck in males are the most frequently cavitated lesions, followed by genitourinary tract lesions in females [35].
2. Infections such as tuberculosis, fungal lesions, pyogenic abscesses, Legionnaires' pneumonia, and parasitic infections.

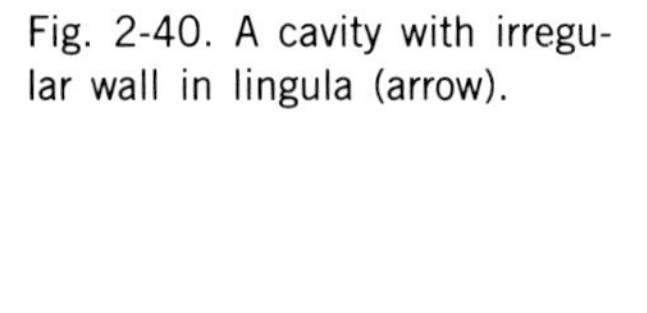

Fig. 2-40. A cavity with irregular wall in lingula (arrow).

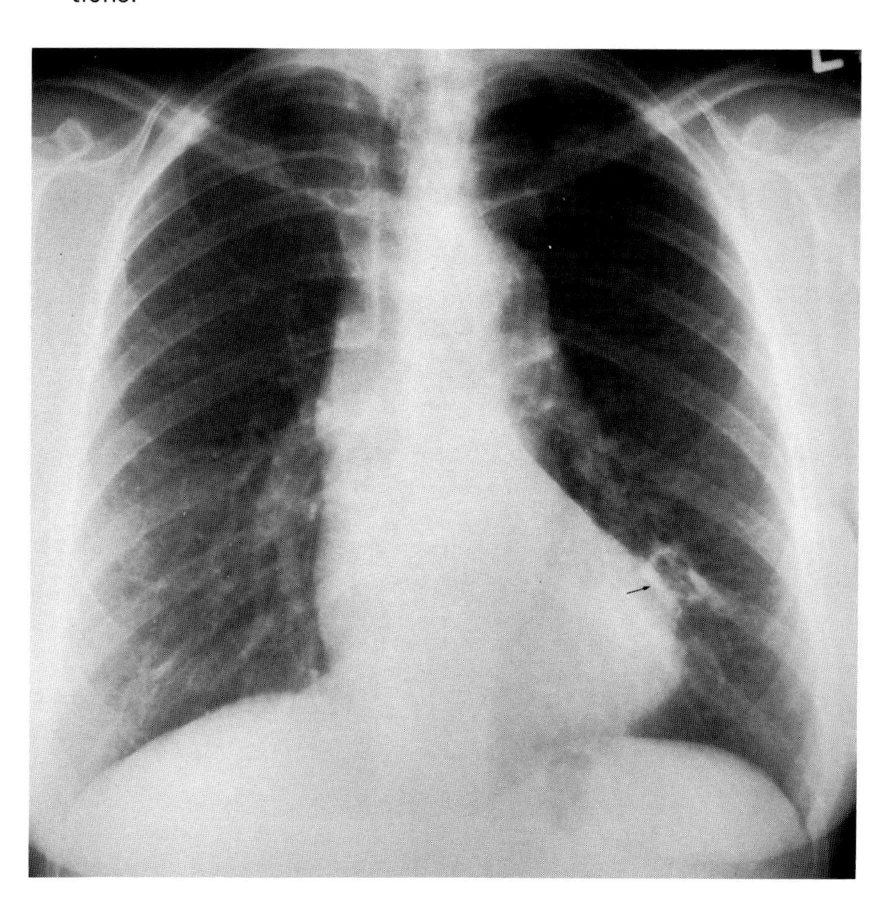

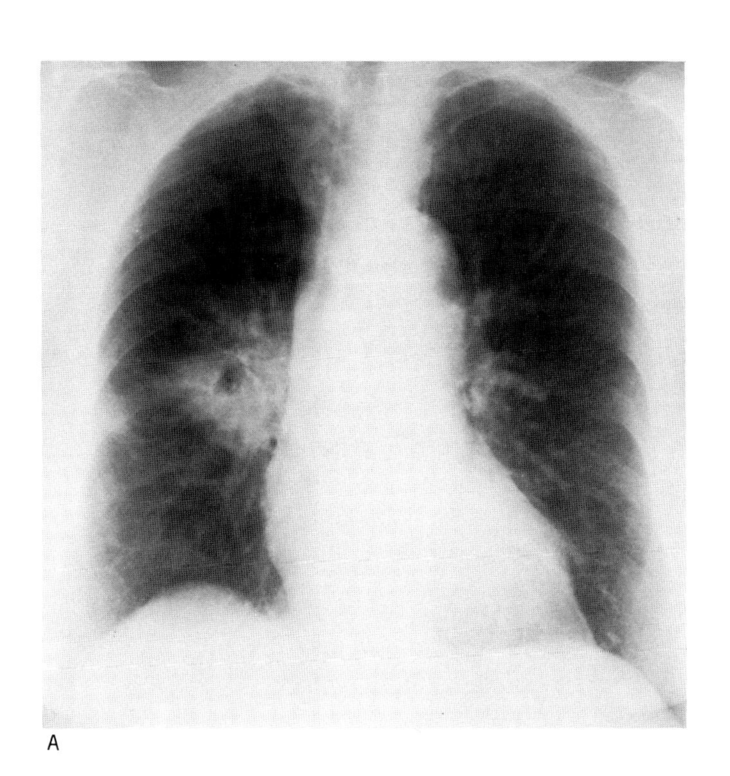

A

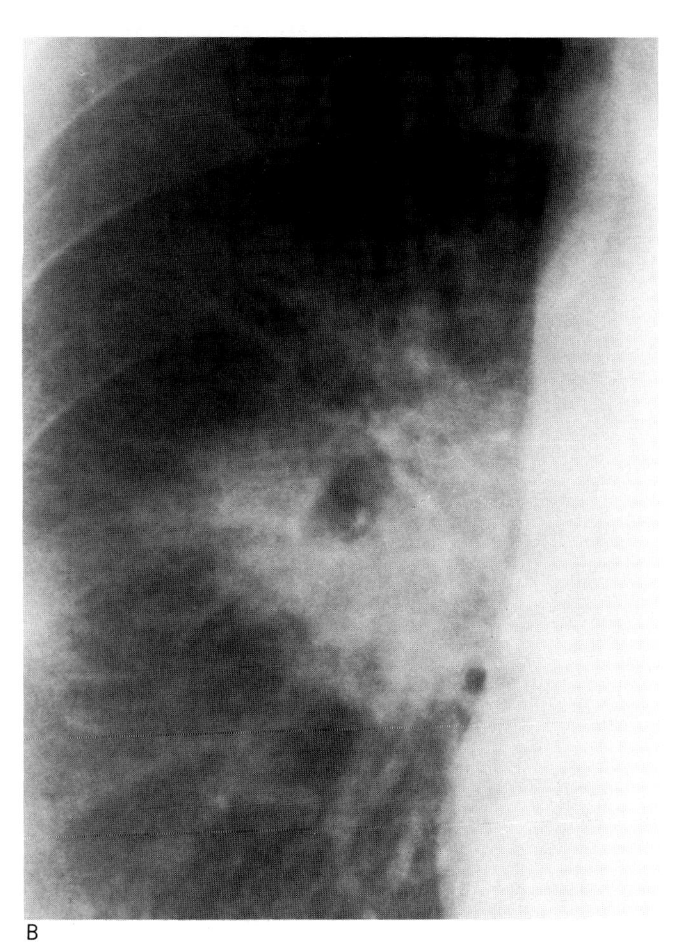

B

Fig. 2-41. *A.* Squamous cell bronchogenic carcinoma showing thick, irregular wall. *B.* Close-up of *A.* *C.* Multiple cavitated septic emboli (arrowheads).

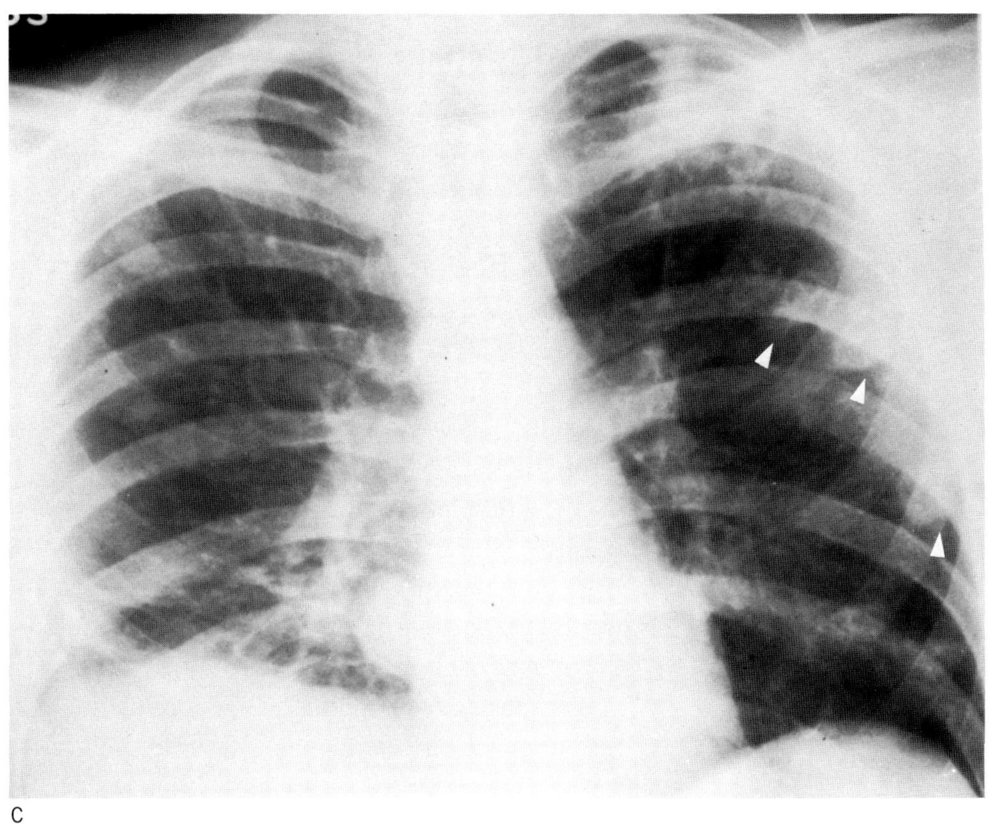

C

3. Autoimmune diseases such as Wegener's granulomatosis, rheumatoid necrobiotic nodule, and polyarteritis nodosa.
4. Diseases of unknown etiology such as sarcoidosis.
5. Infarcts and septic emboli.

Thin-walled cavities and cysts are seen in the following conditions:

1. Congenital cysts, as in cystic adenomatoid malformation and congenital cystic bronchiectasis.
2. Blebs, bullae, and bronchiectatic lesions.
3. Improving infections such as coccidiodomycosis and pneumatocele formation in staphylococcal pneumonia.
4. Traumatic lung cyst and posthydrocarbon ingestion pneumatocele (Fig. 2-42).

A mass or tissue other than just fluid seen within a cavity is found in the following:

1. Fungal infections (such as aspergillosis with a fungus ball [Fig. 2-43]).
2. Tuberculosis with a Rasmussen's aneurysm.
3. Parasitic infection (echinococcal cyst).
4. Carcinoma, bleb carcinoma.
5. Abscess.

PULMONARY INFARCTION

An area of consolidation may represent an infarcted segment of lung following an embolic episode. It may be accompanied by changes secondary to volume loss, such as elevation of the diaphragm. The density has a segmental distribution, is homogeneous in appearance, is often pleura based (Fig. 2-44), and may show an increase in size with sudden tapering of the feeder artery. A Hampton's hump (a wedge-shaped density with a pleural base and a rounded, convex apex toward the hilum) is a typical configuration of an infarct [36]. It rarely shows an air-bronchogram or cavitation. Infarction can occur within 10 hours and up to 1 week after the embolic episode. The resolution of an infarct varies; resolution occurs in 4 to 7 days if only hemorrhage or edema has occurred in the infarcted area [37], and in 20 days to 5 weeks if necrosis has occurred [38].

Woesner and associates [39] reported that the pattern of resolution of an infarct helps to differentiate it from a pneumonic consolidation. Pneumonic shadows break up into areas

Fig. 2-42. *A.* Infiltrate from kerosene aspiration in the right middle lobe. *B.* In patient from *A,* arrows outline the thin walls of the pneumatocele that developed in the previous area of infiltration.

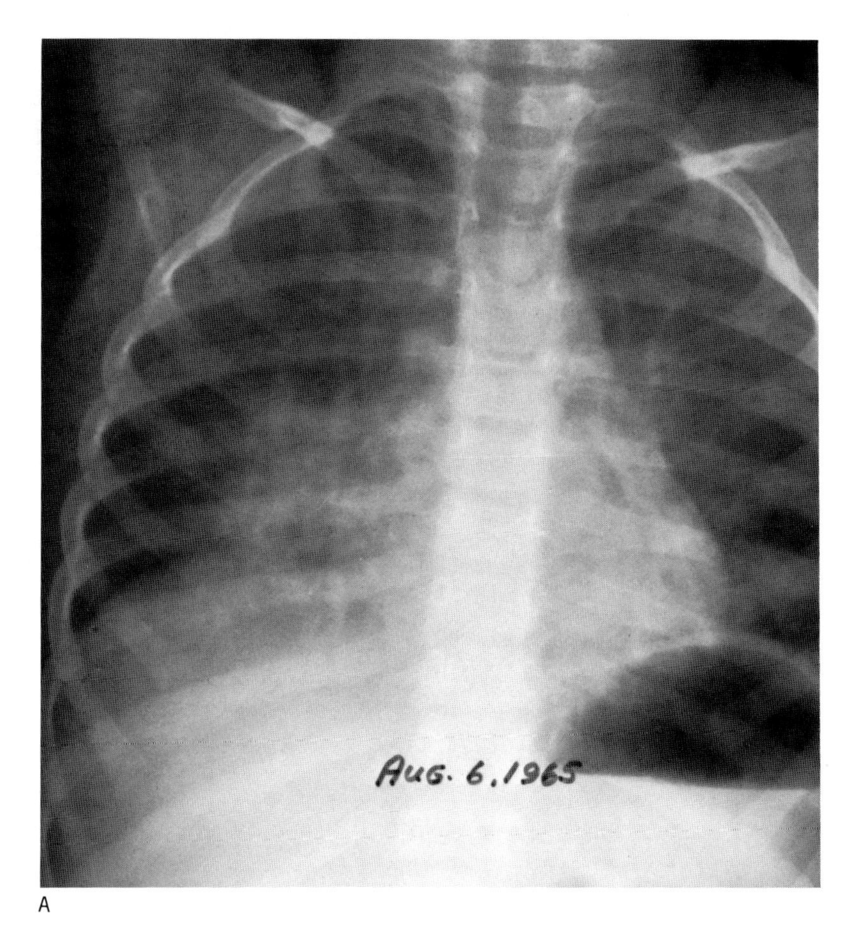

A

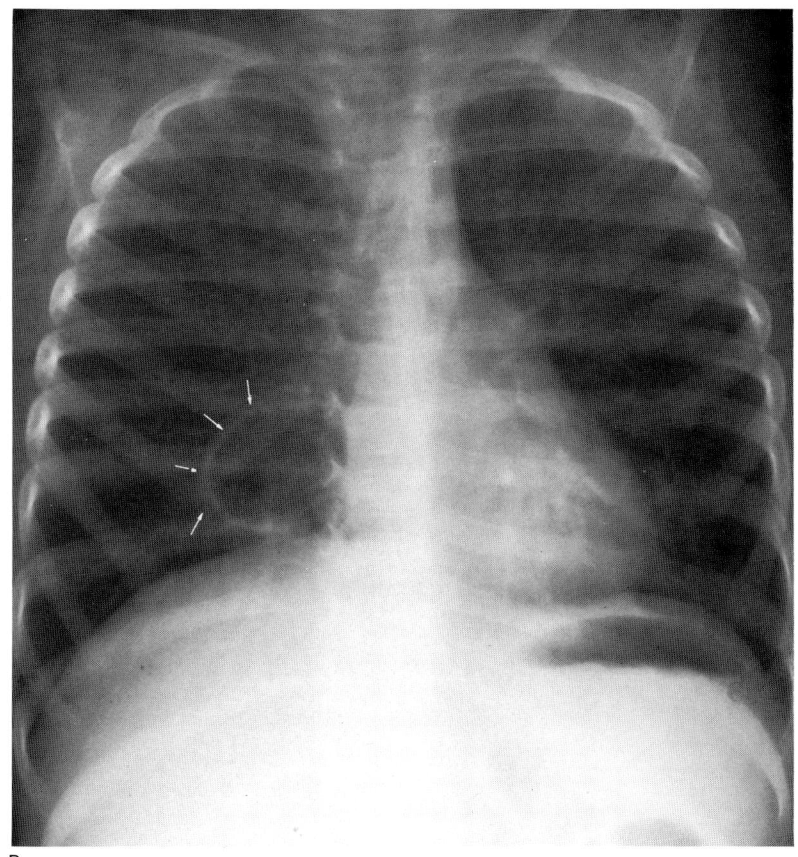

B

Fig. 2-43. *A. Aspergillus* fungus balls seen as soft tissue densities (arrows) within cavities. *B.* Close-up of the right upper lobe of the patient in *A.*

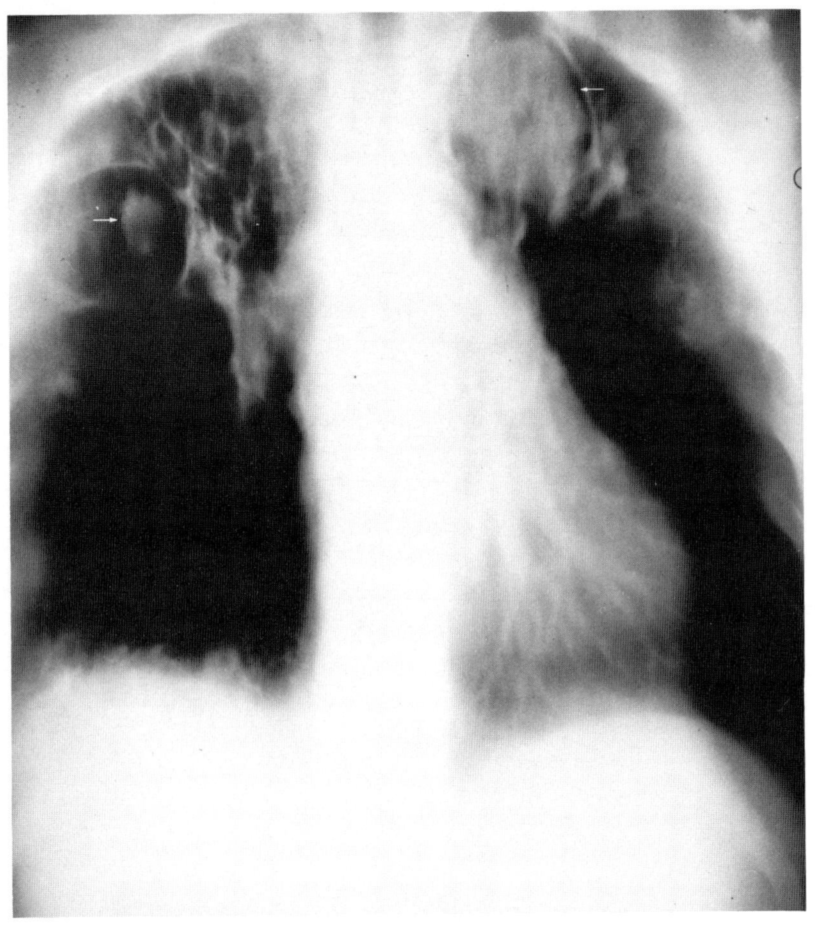

A

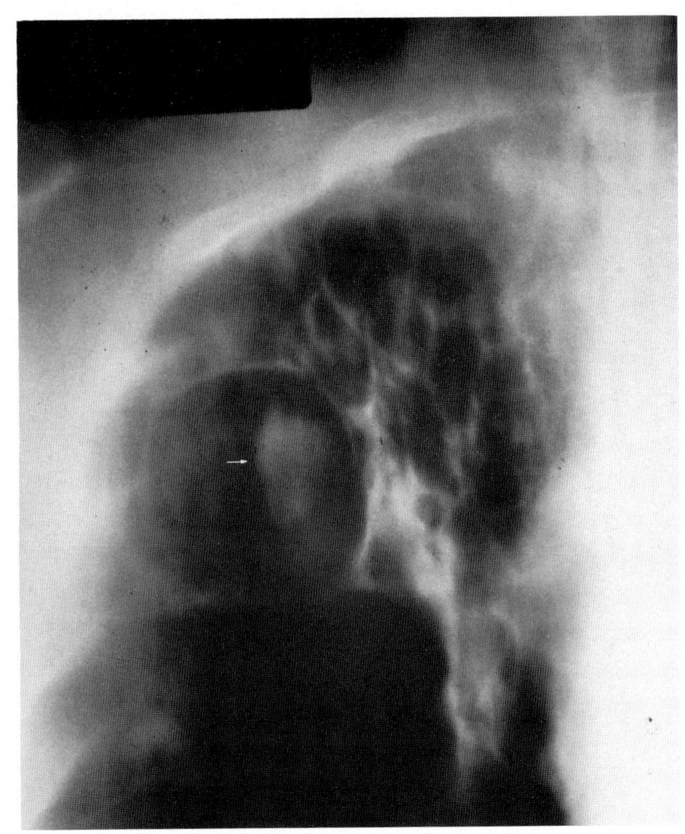

B

Fig. 2-44. *A.* Right pleural effusion with the top of the lower lobe density (arrow) visible above the fluid level. *B.* Ten days later, after the resolution of the effusion, the infarcted pulmonary segment that has undergone partial resolution (arrow) at its periphery is seen. *C.* Lateral view of the same patient. Open arrow points to the Hampton's hump. Note that the lesion is pleura based.

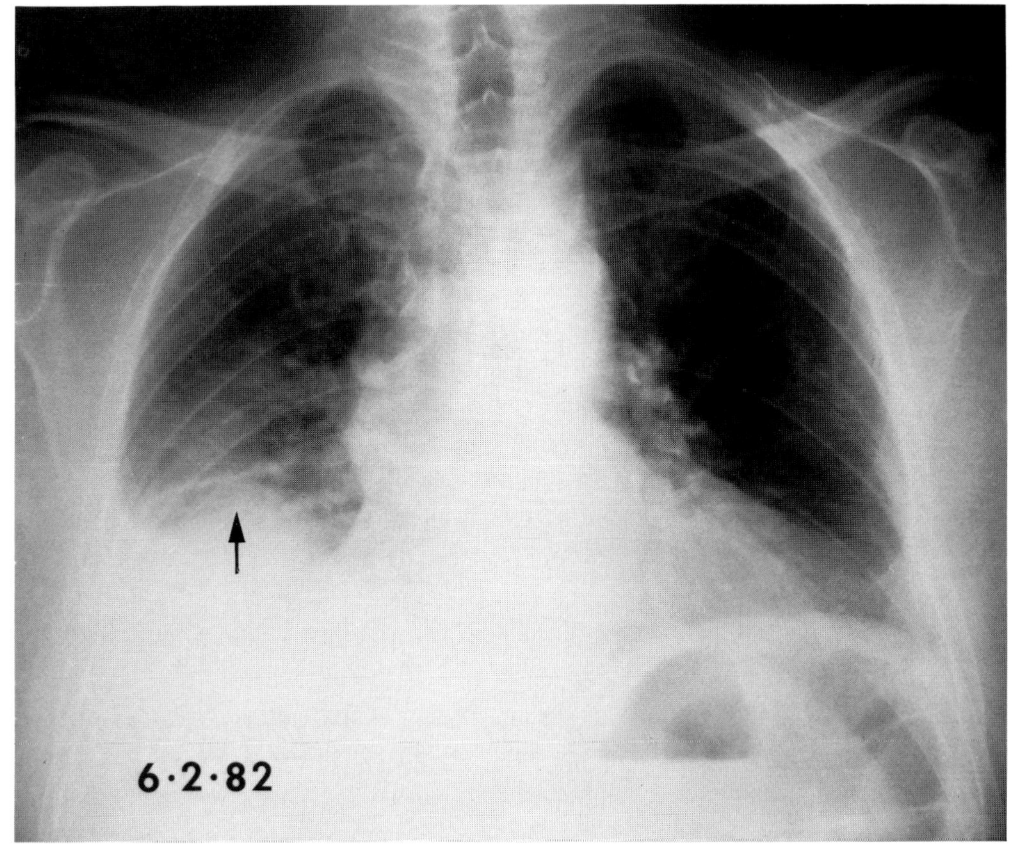

A

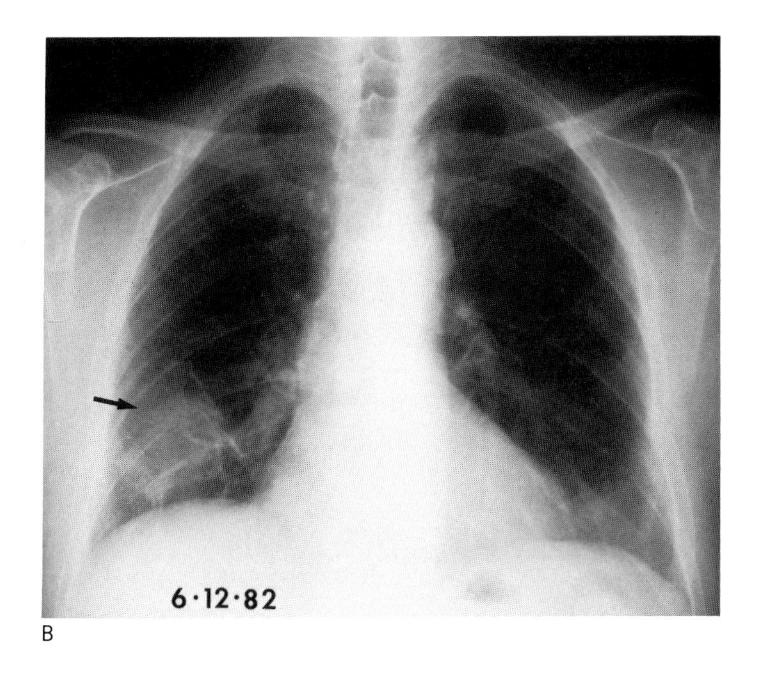

B

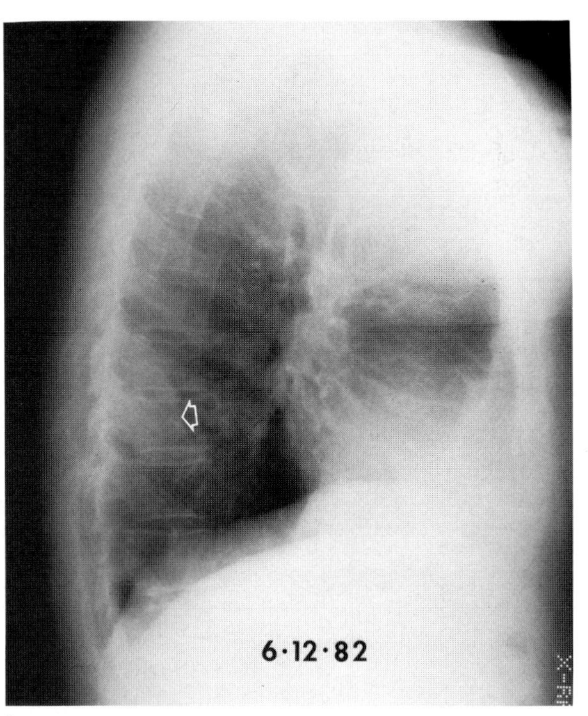

C

of inhomogeneous densities as areas of lucency appear within them; infarcts, on the other hand, diminish in size but maintain their density and shape.

We must point out that most thromboembolic episodes do not result in an infarct. In fact, most episodes show no radiographically detectable changes. The other radiographic changes, aside from infarction, that may be seen in pulmonary embolism are the following:

1. Line shadows. These may be secondary to atelectasis or Fleischner's lines. Fleischner's lines may represent scarring secondary to necrosis, thrombosed vessels (often veins), or thickened pleura.
2. Changes suggesting volume loss, such as elevation of the diaphragm.
3. Pleural effusion.
4. Local area of oligemia (Westermark's sign) [40] secondary to contusion of a segmental artery and spasm of the smaller vessels. Localized oligemia is often seen during the first days after the episode and may disappear within 24 to 36 hours.
5. Generalized oligemia resulting from widespread occlusion of smaller arteries. It is often accompanied by changes of pulmonary hypertension.
6. Enlargement of the major hilar arteries secondary to distension of the artery by the bulk of the thrombus [41] (see Hilar Vasculature in Step 3 for measurement of hilar artery size). The increased size of the vessel accompanied by sudden tapering is called a *knuckle sign* [42,43,44] (see Fig. 3-8).
7. Occasionally, changes of cor pulmonale.

INTERSTITIAL DISEASE

Interstitial disease produces four basic roentgenographic patterns [41]:

1. Ground glass pattern, appearing as a relatively homogeneous haze or clouding of the lungs.
2. Nodular pattern, consisting of tiny interstitial nodules, as in miliary tuberculosis (see Fig. 2-29) and early disseminated carcinomatosis.
3. Reticular pattern (Fig. 2-45), consisting of a network of linear, ringlike opacity surrounding air-filled spaces.
4. Reticulonodular pattern (Fig. 2-46), which is an admixture pattern.

Fig. 2-45. *A.* Bilateral reticular densities in a patient with congestive heart failure. *B.* Bilateral reticular infiltrates in a patient with early *Pneumocystis carinii* pneumonia.

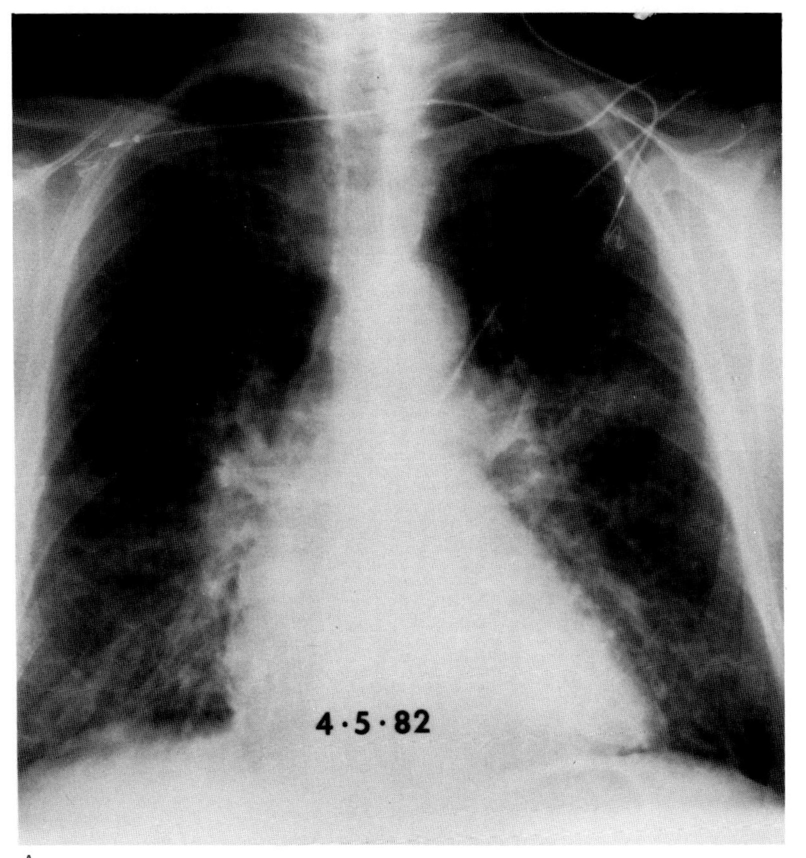

A

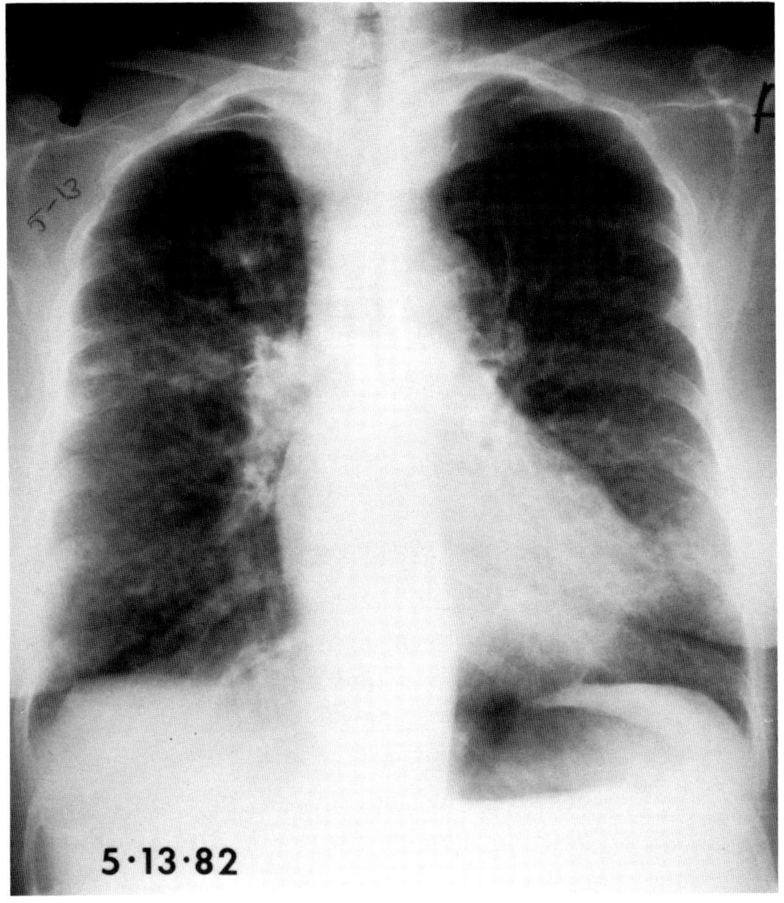

B

In most lung diseases, the changes are usually a combination of consolidation and atelectasis or interstitial infiltrates or all three. Interstitial infiltrates are often recognized as linear densities, often reticular and separate from the normal vascular markings. They may be admixed with interstitial nodules. Enlargement or thickening of the intralobular septae produces Kerley B lines [45] (short linear densities perpendicular to and continuous with the pleura) (Fig. 2-47A) and Kerley A lines [46] (longer linear strands crossing vascular markings and radiating to the hila) (Fig. 2-47B).

To remember the numerous causes of interstitial diseases one can break them down according to etiology. Table 2-3 lists this breakdown.

The mnemonic *hide facts,* taken from the first letters of the common diseases, can be used to develop a differential diagnosis. The letters stand for the following:

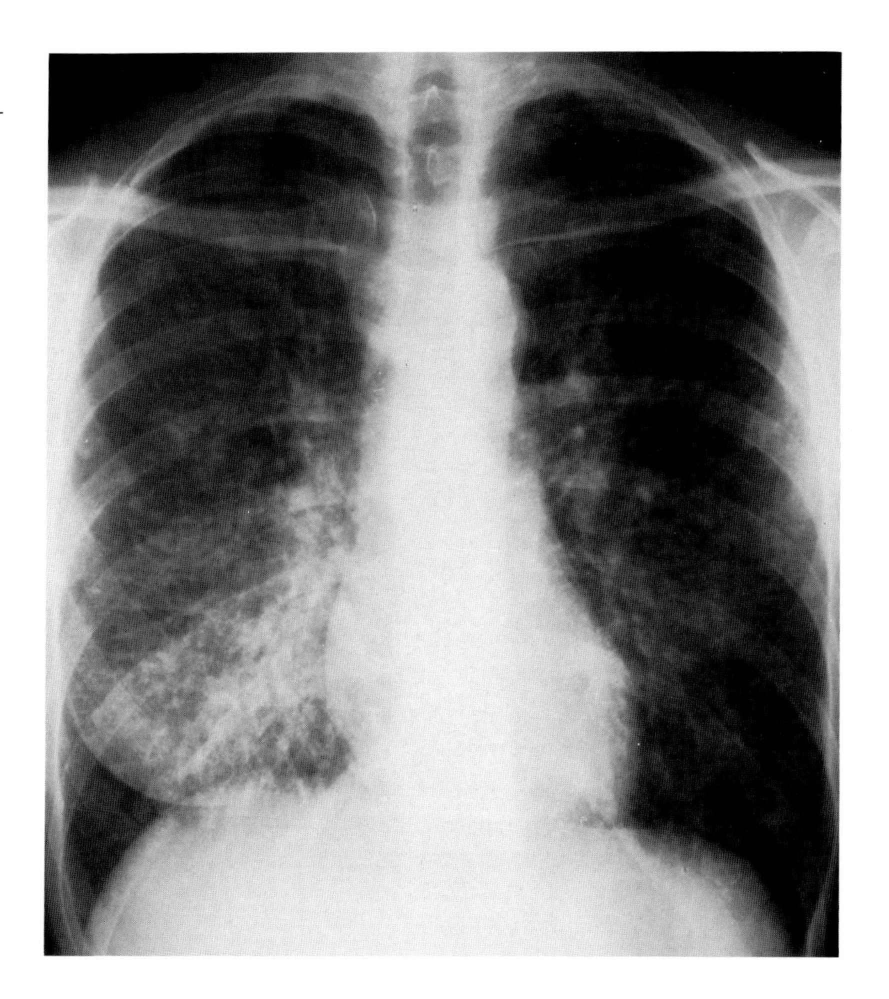

Fig. 2-46. Reticulonodular pattern in a patient with lymphangitic metastasis from breast carcinoma. Note absence of left breast.

Histoplasmosis and hypersensitivity pneumonitis
Interstitial pneumonias (UIP, DIP, LIP) and idiopathic fibrosis
Drug-induced disease
Eosinophilic granuloma
Fungal disease
Asbestosis and other inhalational diseases
Collagen vascular diseases (rheumatoid, sclerodermas)
Tuberculosis and tuberous sclerosis
Sarcoidosis

TABLE 2-3. Commonly Considered Causes of Interstitial Diseases

---

Congenital conditions
- Tuberous sclerosis
- Neurofibromatosis

Infections
- Acute
  - Viral
  - Mycoplasma
  - Infectious mononucleosis
  - *Pneumocystis carinii*
- Chronic
  - Tuberculosis
  - Fungal infections

Pulmonary edema
- Congestive failure
- Renal failure
- Mitral stenosis

Neoplastic processes
- Lymphangitic metastasis
- Lymphoma
- Leukemia

Inhalational diseases
- Pneumoconiosis
- Allergic alveolitis

Collagen vascular diseases
- Rheumatoid arthritis
- Scleroderma
- Lupus erythematosus

Iatrogenic diseases
- Drug induced, as from bleomycin and other antineoplastic chemotherapeutic agents
- Radiation induced

Idiopathic sources
- Eosinophilic granuloma
- Interstitial pneumonias (UIP, DIP, LIP, PIP)
- Alveolar proteinosis
- Idiopathic pulmonary hemosiderosis
- Sarcoid

A

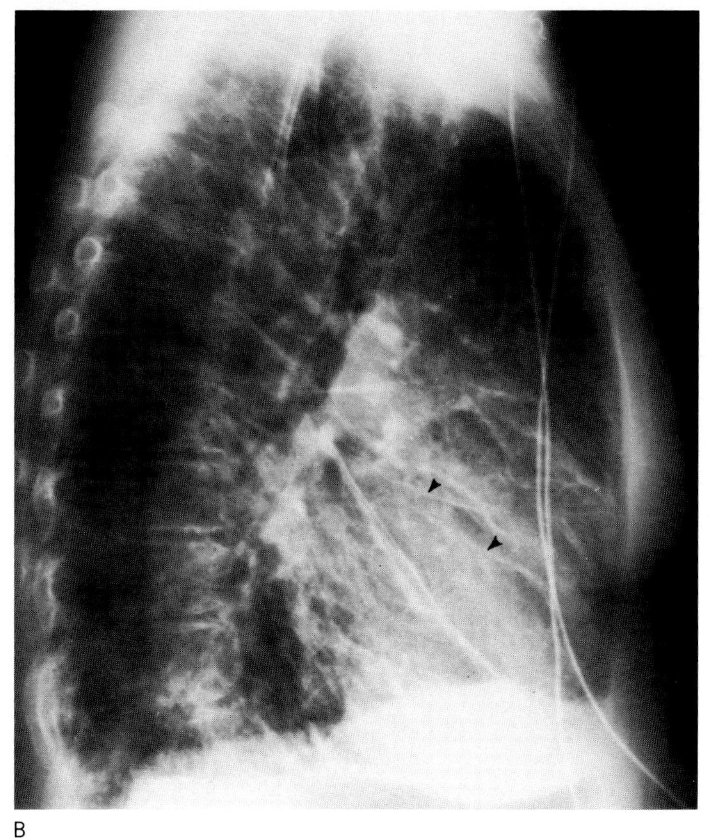

B

Fig. 2-47. *A.* Kerley B lines (arrowheads). *B.* Kerley A lines (arrowheads).

Some of these interstitial diseases cause honeycombing. Honeycombing is seen in end-stage lung disease as cystic spaces, 5 to 10 mm in diameter, with thickened fibrous walls and distorted intervening parenchyma. To remember which of the interstitial diseases cause honeycombing, one can use the mnemonic *fairest* and the phrase *the fairest of them all produce the honeycombs.* The letters stand for the following:

Fungal disease (chronic interstitial disease)
Asbestosis
Idiopathic fibrosis
Rheumatoid arthritis
Eosinophilic granuloma
Scleroderma
Tuberous sclerosis

Interstitial diseases are known to have anatomic predominance, which is extremely useful in preparing a differential diagnosis.

Fig. 2-48. Bilateral, predominantly basal, interstitial fibrosis with pleural calcification in both upper hemithoraces laterally in a patient with asbestosis.

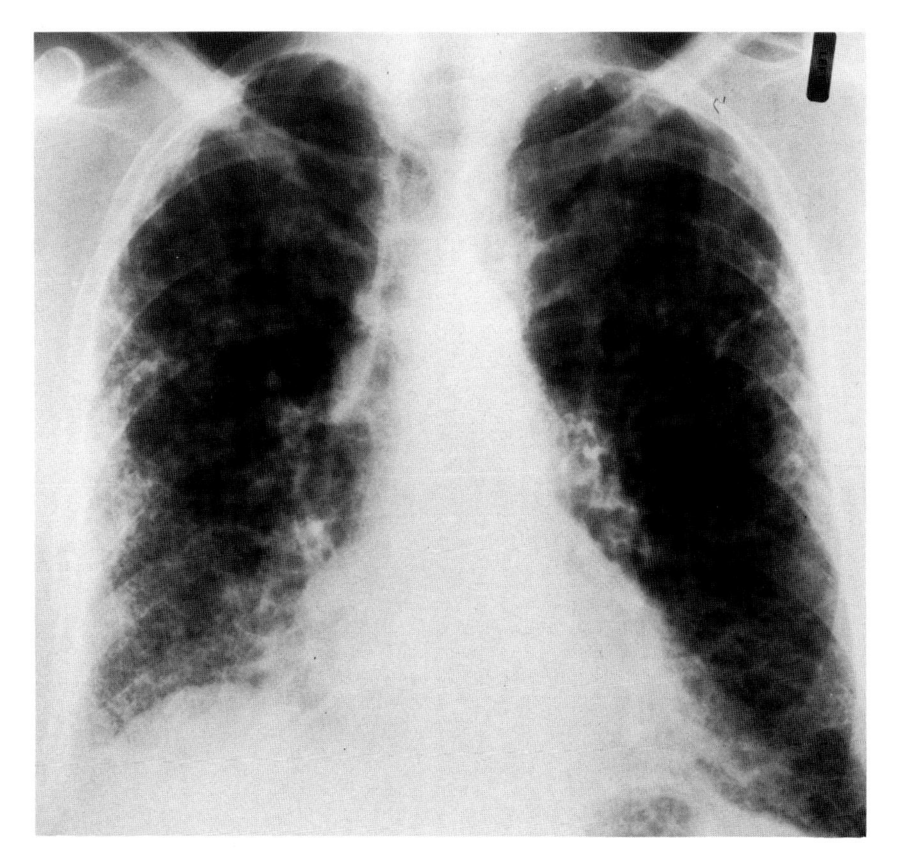

Upper lobe predominance
- Silicosis
- Histiocytosis
- Early stages of sarcoidosis
- Tuberculosis
- Cystic fibrosis

Lower lobe predominance
- Asbestosis (Fig. 2-48)
- Collagen vascular disease
- Drug allergy
- Idiopathic fibrosis

HYPERINFLATION

Hyperinflation refers to overdistended alveoli, whereas emphysema (Fig. 2-49) denotes destruction of the alveolar septae and air spaces distal to the terminal bronchiole. Hyperinflation may occur as a compensatory mechanism whenever volume loss occurs through atelectasis or surgical removal of lung tissue. Hyperinflation also occurs with a check-valve obstruction that allows air to be drawn in during inspiration but prevents egress on expiration. Fluoroscopic examination or a

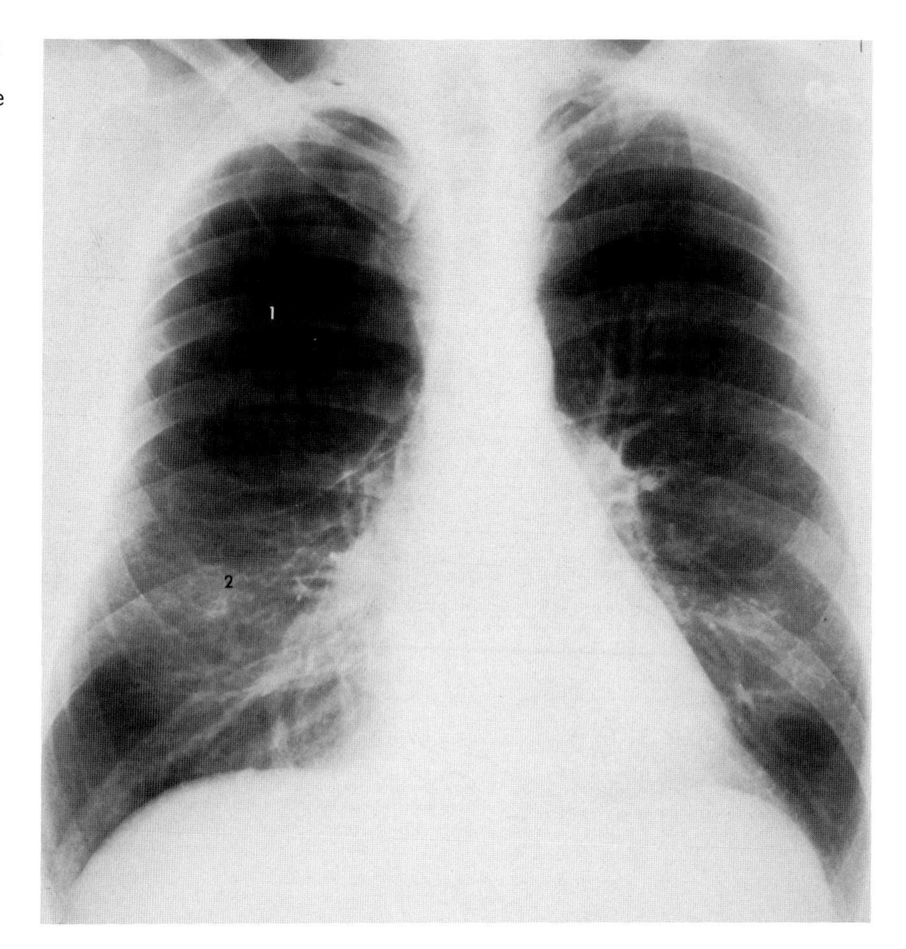

Fig. 2-49. Emphysematous right upper lobe (area 1) appears more lucent than left upper lobe and has few stringy lung markings. Area 2 represents lung compressed by the emphysematous upper lobe. Note crowding of lung markings in the compressed lung.

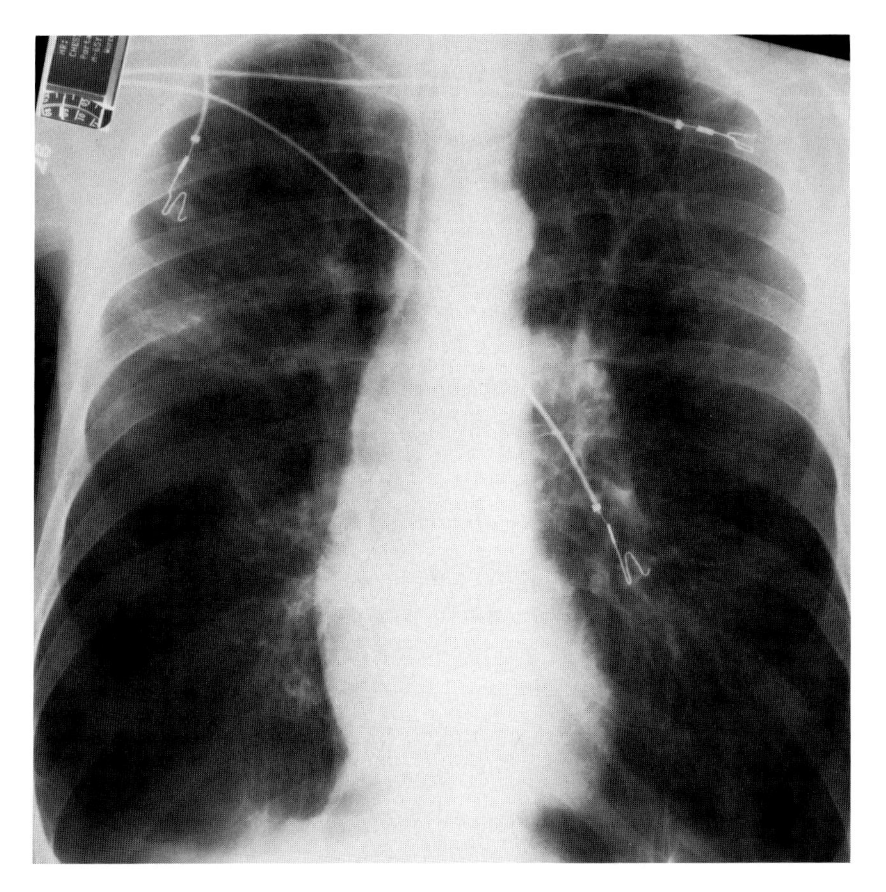

Fig. 2-50. Marked emphysema. Note nonuniform distribution of vascular markings, increased radiolucency of the bases, inverted diaphragm, and a small vertical heart.

combination of inspiration-expiration chest films will show mediastinal shift away from the hyperlucent side on expiration. These procedures are necessary to make the diagnosis of check-valve obstruction.

Emphysema, on the other hand, shows changes in the lung, heart, great vessels, diaphragm, and thoracic cage (Fig. 2-50). It is seen as increased radiolucency of the lungs associated with the presence of blebs and bullae. The depth of the retrosternal air space increases, and the anteroposterior diameter of the chest increases, while the diaphragm flattens, or may even become inverted. If fluoroscopy is performed, the normal 5 to 10 cm excursion of the diaphragm from inspiration to expiration is reduced to less than 3 cm.

Both hyperinflation and emphysema cause the heart to appear small and vertical. In emphysema, the central vessels are prominent, and the peripheral vessels show a nonuniform distribution, because blood has been diverted away from the destroyed lung to the normal areas. On the other hand, a different pattern of emphysema, increased-markings emphysema, is manifested by mild to moderate hyperaeration and an increase in the lung markings. There is often associated cardiomegaly and prominence of the central vessels. This form of emphysema is seen in patients with chronic bronchitis.

STEP 3 EVALUATE THE VASCULAR STRUCTURES

HILAR VASCULATURE

The right hilum forms a *V* lying on its side (>), with the upper limb formed by the right superior pulmonary vein and the lower limb formed by the right interlobar pulmonary artery (Fig. 3-1). If the angle is obliterated or becomes convex, a hilar mass (Fig. 3-2) or pulmonary vascular congestion (Fig. 3-3) should be suspected. The left side is more difficult to evaluate, so familiarity with its normal appearance is important in order to distinguish minimal abnormality. Often the left hilum is seen in the shape of a walking cane (Fig. 3-4). Vessels in both hila have relatively straight borders that follow the bronchi. The presence of lobulation, except where vessels cross one another, usually indicates disease (Fig. 3-5).

In most people, the right hilum is 0.75 to 3.0 cm lower than the left. In 3 percent of the population, the two are at the same level; in no instance is the right hilum higher than the left [9]. As the pulmonary vessels subdivide, the transverse diameter of a branch should be approximately half that of the more proximal or more medial artery. A sudden change in caliber suggests pulmonary arterial hypertension (Fig. 3-6). The right interlobar (descending) pulmonary artery's maximum normal transverse diameter is 16 mm [47], measured at the level of the bronchus intermedius (Fig. 3-7). A diameter greater than this in a patient suspected of having a pulmonary embolus should be considered pathologic (Fig. 3-8).

On the lateral projection, the right pulmonary artery projects anterior to the trachea, and the left forms an arch paralleling the inferior aspect of the aortic arch (Fig. 3-9).

The margin of the right pulmonary artery is discrete in only a small portion of the population, because it nearly always divides within the pericardium. The left, because it divides outside the pericardium, commonly enters the lung as a well-defined structure outlined by air. Small hilar areas may be seen in patients with decreased pulmonary vascularity from a right to left shunt, as in tetralogy of Fallot.

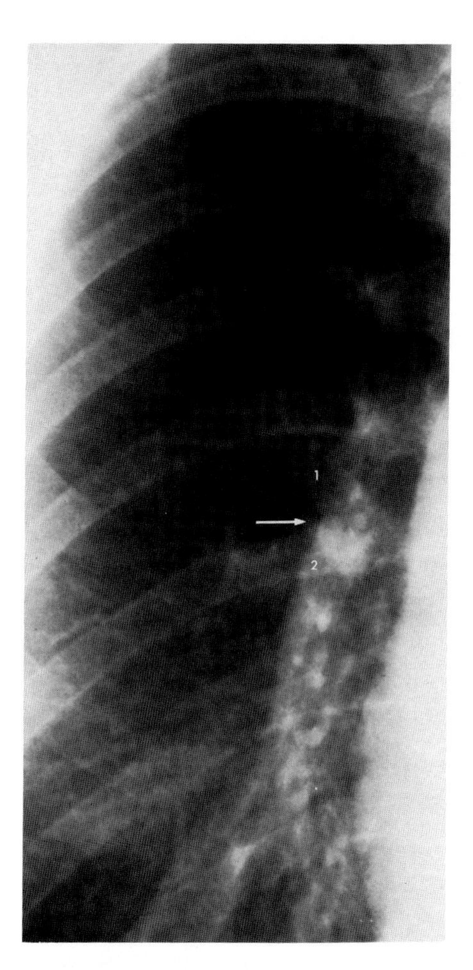

Fig. 3-1. V-shaped hilum. Arrow points to apex of V, 1 to the right superior pulmonary veins, and 2 to the right interlobar pulmonary artery.

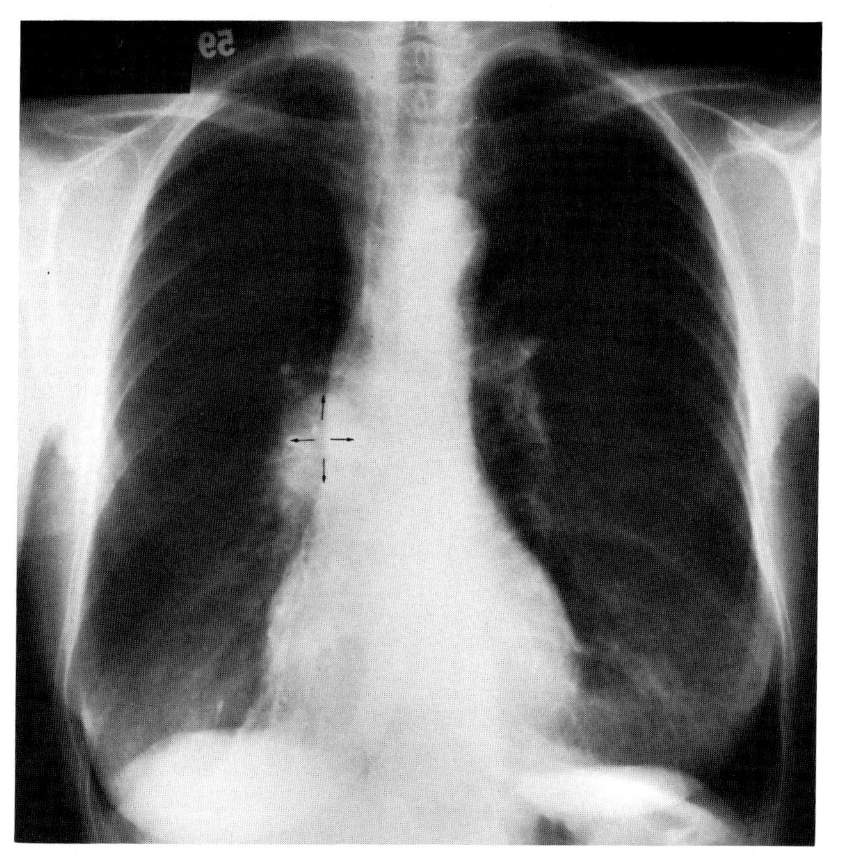

Fig. 3-2. Arrows demonstrate a right hilar mass obliterating the angle of the right hilum and producing an increase in the density of the right hilum.

Fig. 3-3. Obliteration of the hilar angle (arrow) produced by enlarged hilar vessels in congestive heart failure.

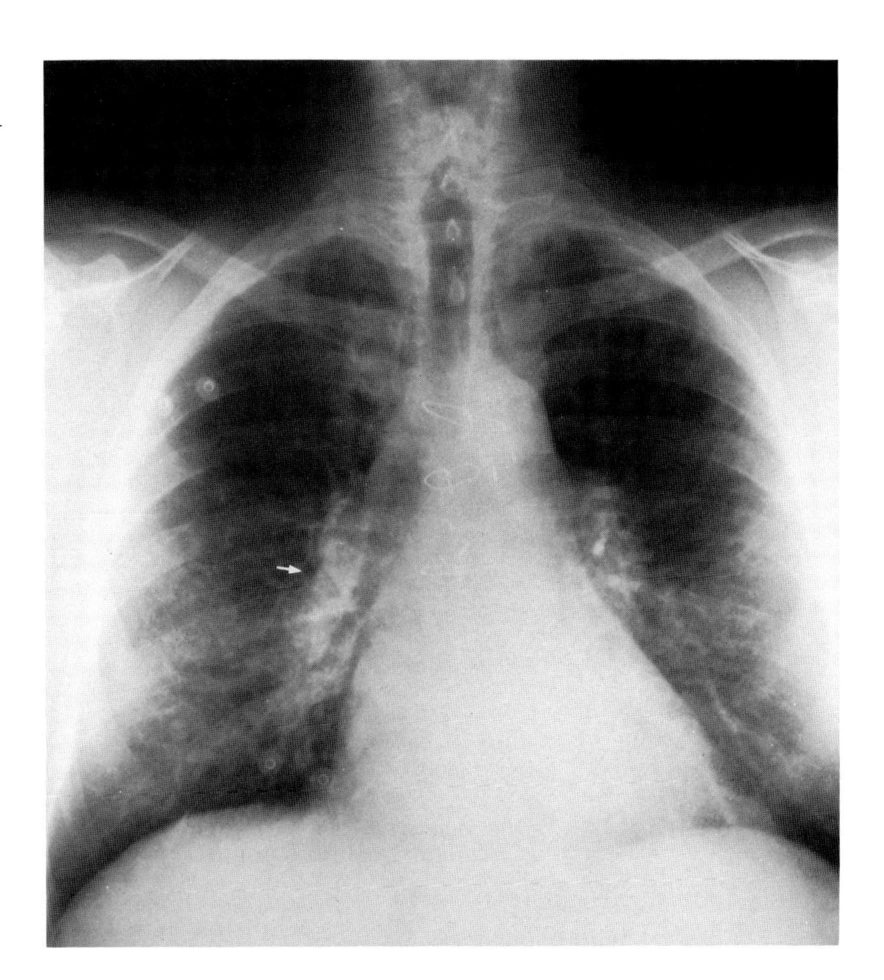

Fig. 3-4. Normal walking cane appearance of the left hilum (traced by arrows) is altered to a small mass (arrowhead) produced by bronchogenic carcinoma.

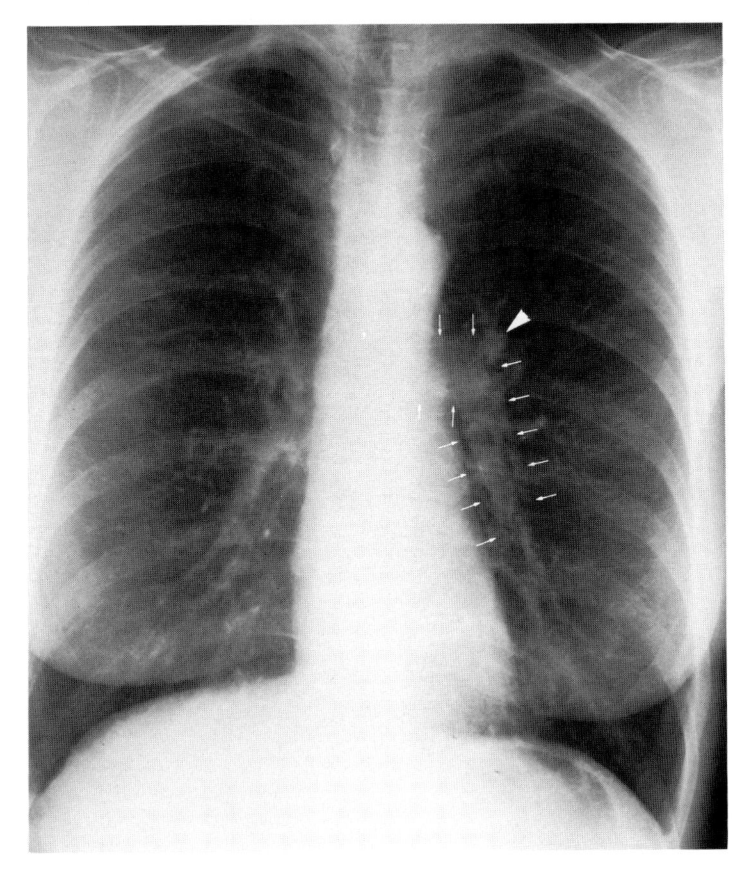

Fig. 3-5. *A*. Large lobulated left hilar masses (arrow) from bronchogenic carcinoma. *B*. Bilateral hilar masses from nodal enlargement secondary to sarcoidosis. *C*. A lobulated density (arrow) in the region of the right pulmonary artery. Density is secondary to a bronchogenic carcinoma.

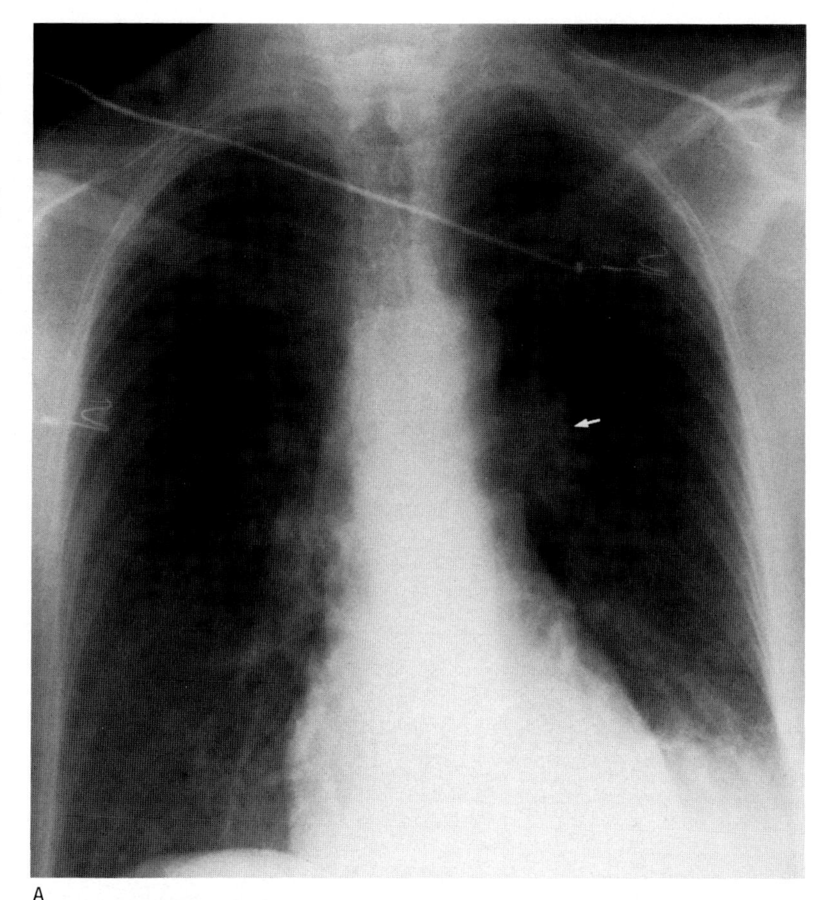

A

B

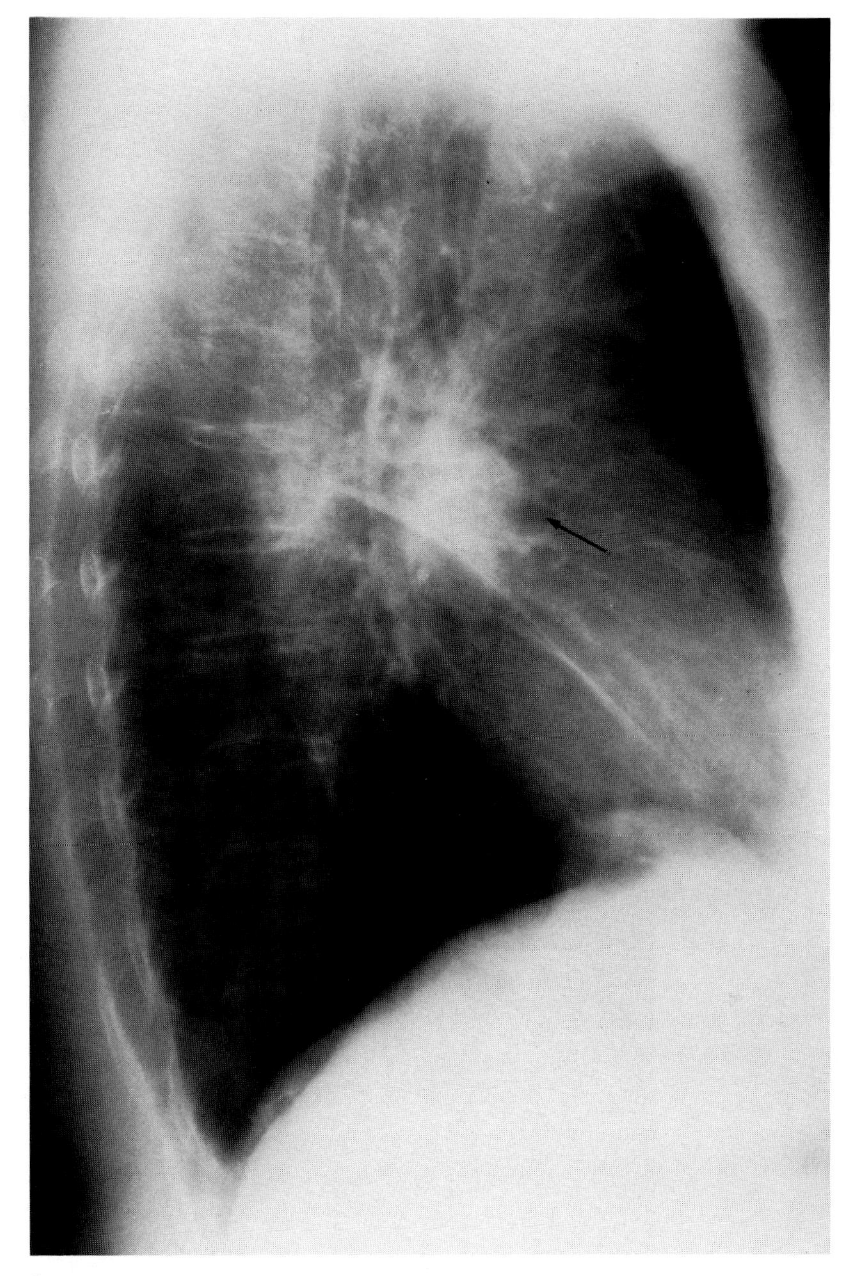

C

Fig. 3-6. *A.* Patient with pulmonary hypertension demonstrating a bulging pulmonary artery segment (arrow) and a marked discrepancy in size between the right interlobar artery (black arrowheads) and its immediate branches (white arrowheads). *B.* Lateral view shows a bulging pulmonary artery (arrows) and an enlarged right ventricle (open arrowhead).

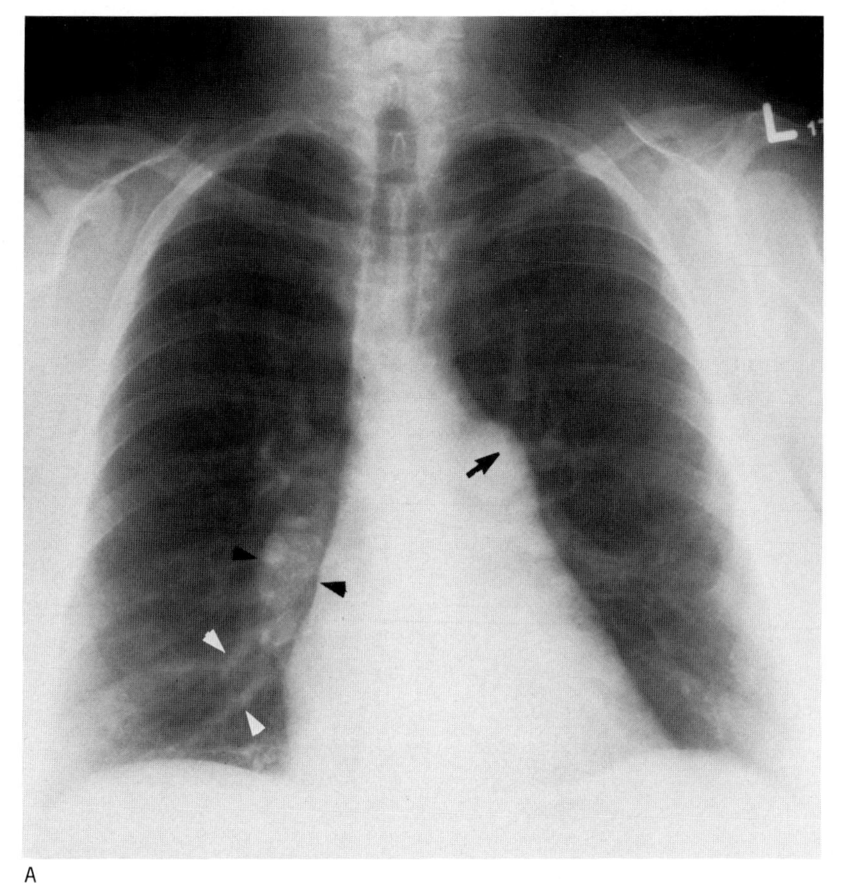

A

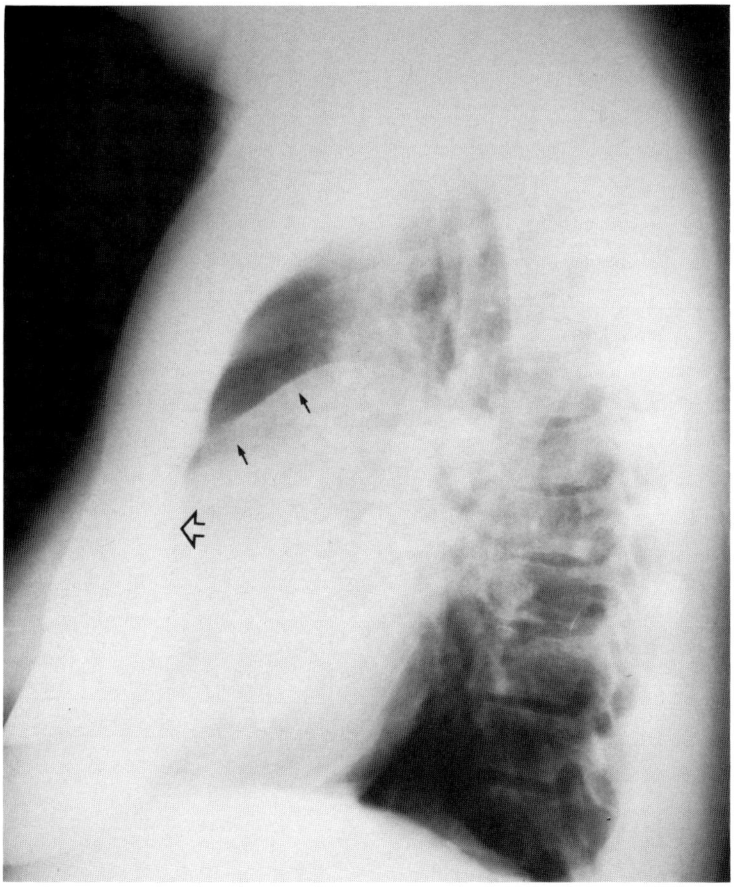

B

Fig. 3-7. Normal right interlobar artery (arrows). Medial arrow located within the bronchus intermedius.

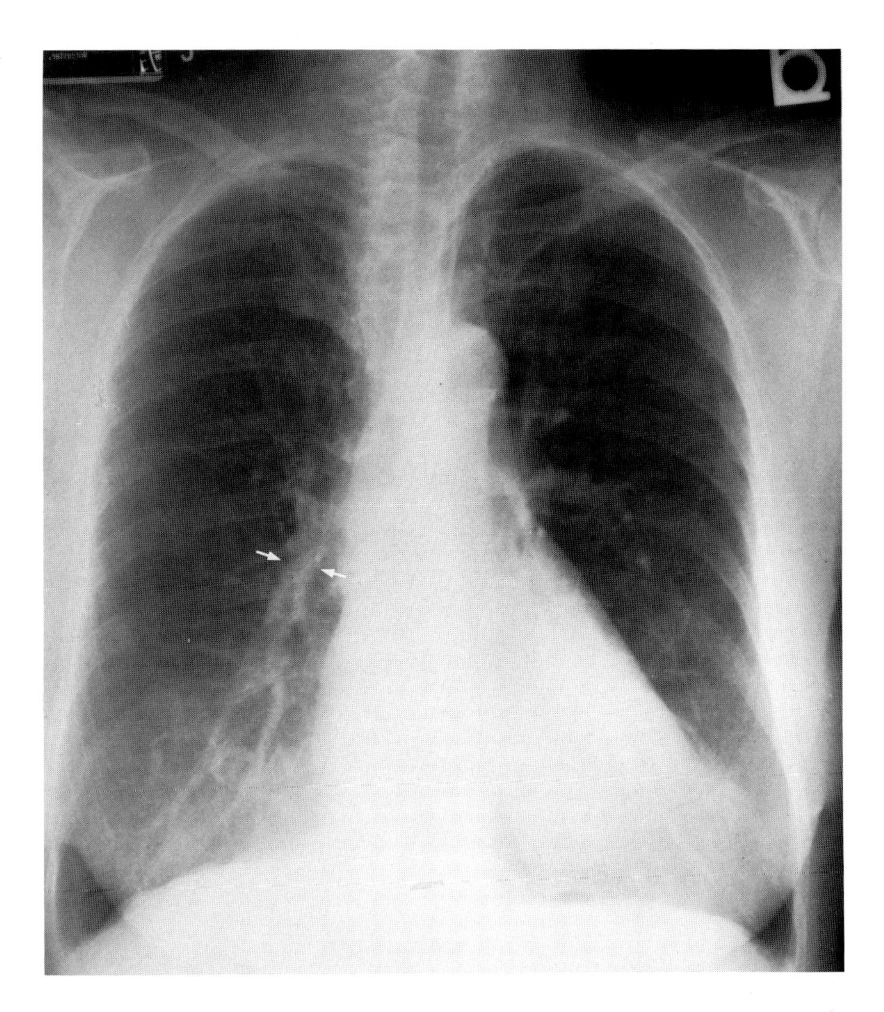

Fig. 3-8. Pulmonary embolus enlarging the interlobar artery (arrow) with an infarct (arrowhead) of a segment of the area supplied by the artery.

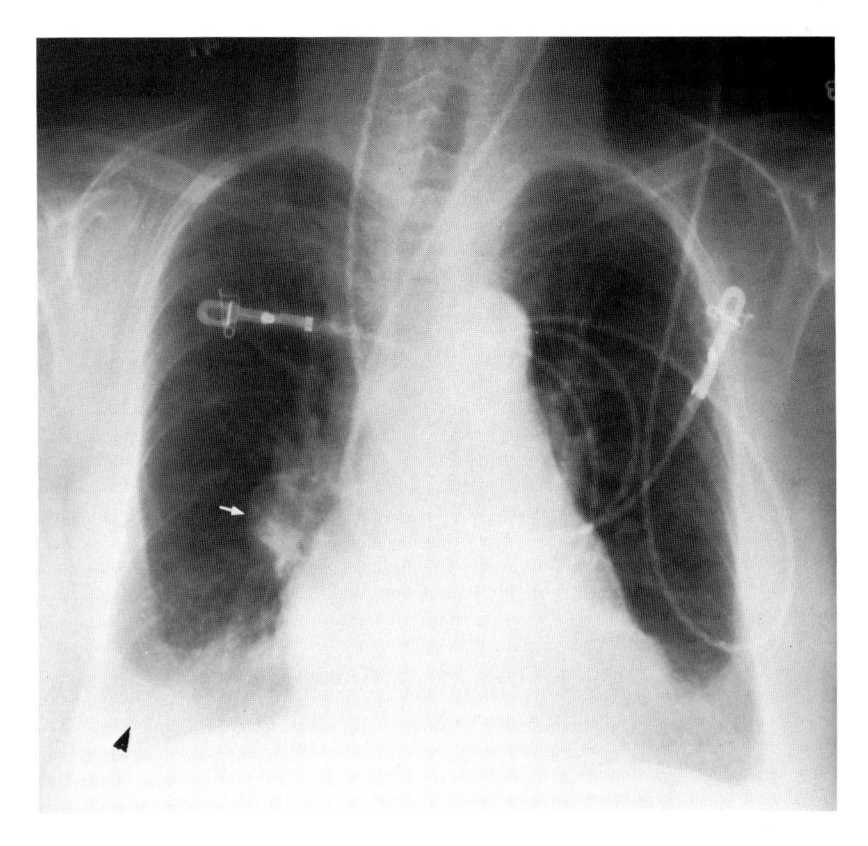

Fig. 3-9. Normal vascular structures in the lateral view: the aortic arch (1), the left pulmonary artery (2), and the right pulmonary artery (3).

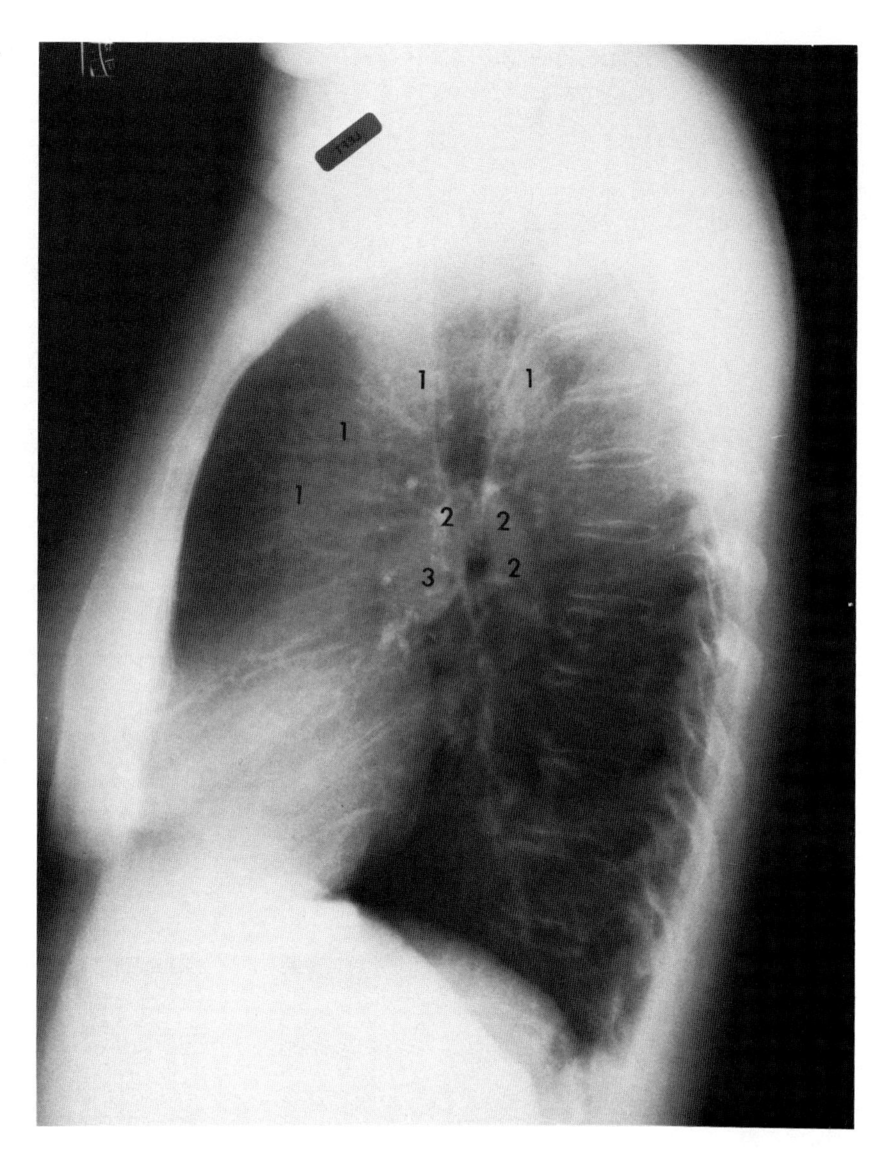

PERIPHERAL PULMONARY VESSELS

To evaluate pulmonary vascularity, one starts by dividing each lung sagittally into three parts in the frontal projection. In the most medial third, one sees the hilar and first-division vessels; in the middle third, the distal pulmonary vessels of smaller caliber; and in the outer third, almost no vessels at all (Fig. 3-10). If vessels extend significantly into the outer third of the lung, one must consider a hemodynamically significant left to right shunt or pulmonary vascular congestion. Diminished peripheral pulmonary vessels can be seen in right to left shunts, pulmonary embolism, and Swyer-James syndrome.

Subsequently, one looks for a "vessel-on-end" adjacent to a "bronchus-on-end." The caliber of the two structures should be almost equal (Fig. 3-11). If the caliber of the vessel is noticeably greater than that of the bronchus, congestion, active or passive, is confirmed (Fig. 3-12).

Fig. 3-10. The right lung has been divided into three zones. Zone 1 contains the big vessels, zone 2 the smaller caliber vessels, and zone 3 is almost devoid of pulmonary vessels. Note also the 2:3 ratio between upper and lower lobe vessels in zone 2.

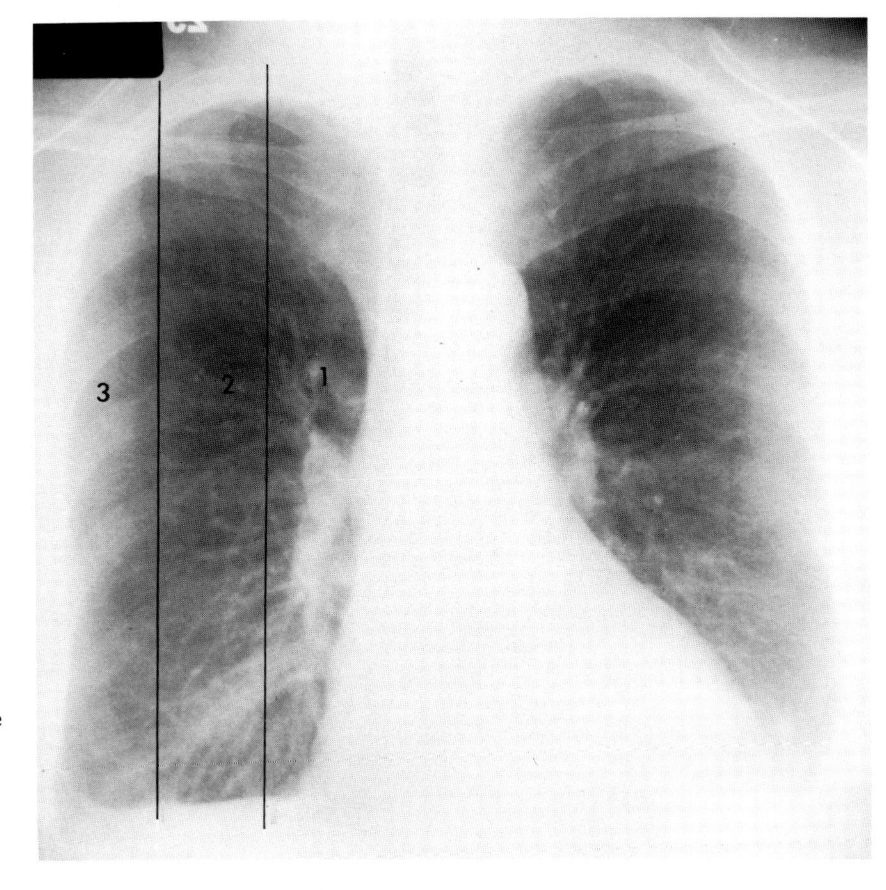

Fig. 3-11. *A.* Close-up of a vessel- (arrow) and bronchus-on-end (arrowhead). Note equal size of the two and the thin bronchial wall. *B.* Close-up of unequal on-end vessel (arrow) and bronchus (arrowhead) in a patient with congestive heart failure.

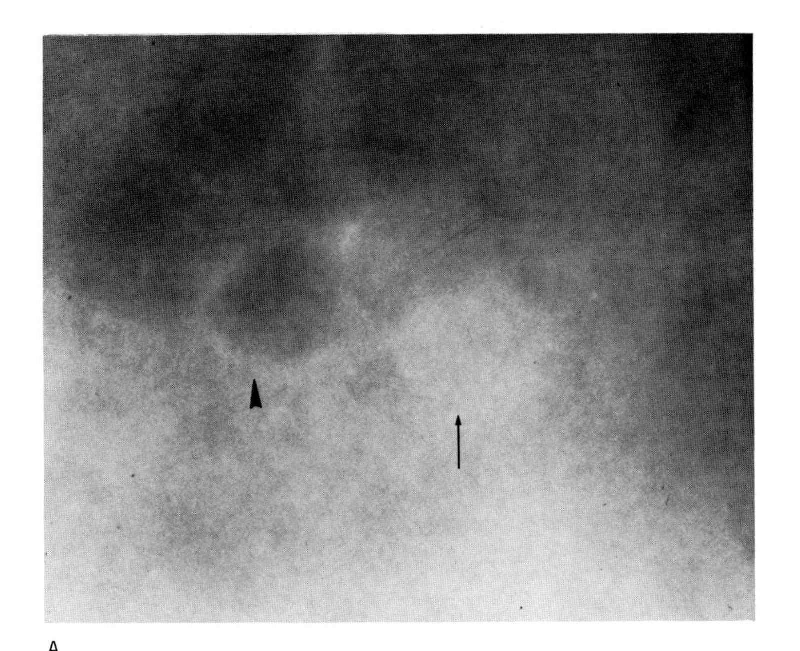

A

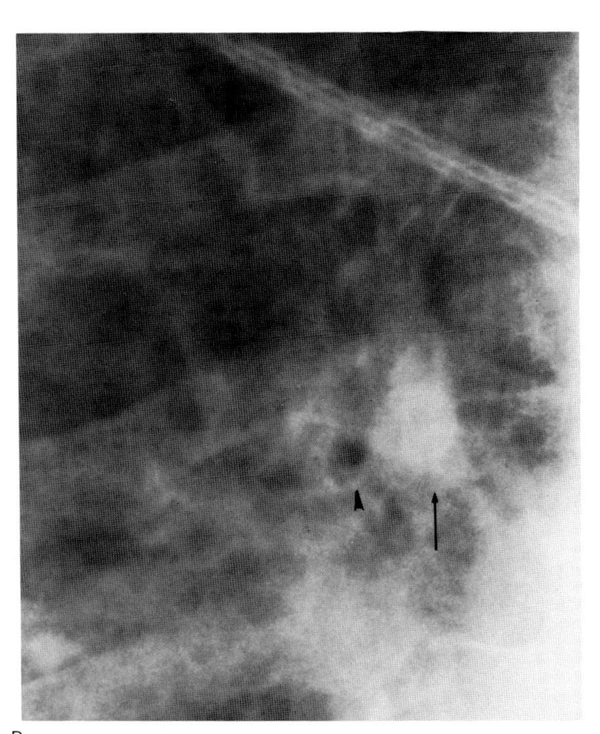

B

Fig. 3-12. *A.* Bronchus (arrowhead) beneath the on-end vessel. Note marked discrepancy in size between the two structures in a patient with congestive heart failure. *B.* Close-up of vessel and bronchus in *A. C.* Return to almost equal size of the vessel and bronchus (arrowhead) with improvement of the congestive heart failure.

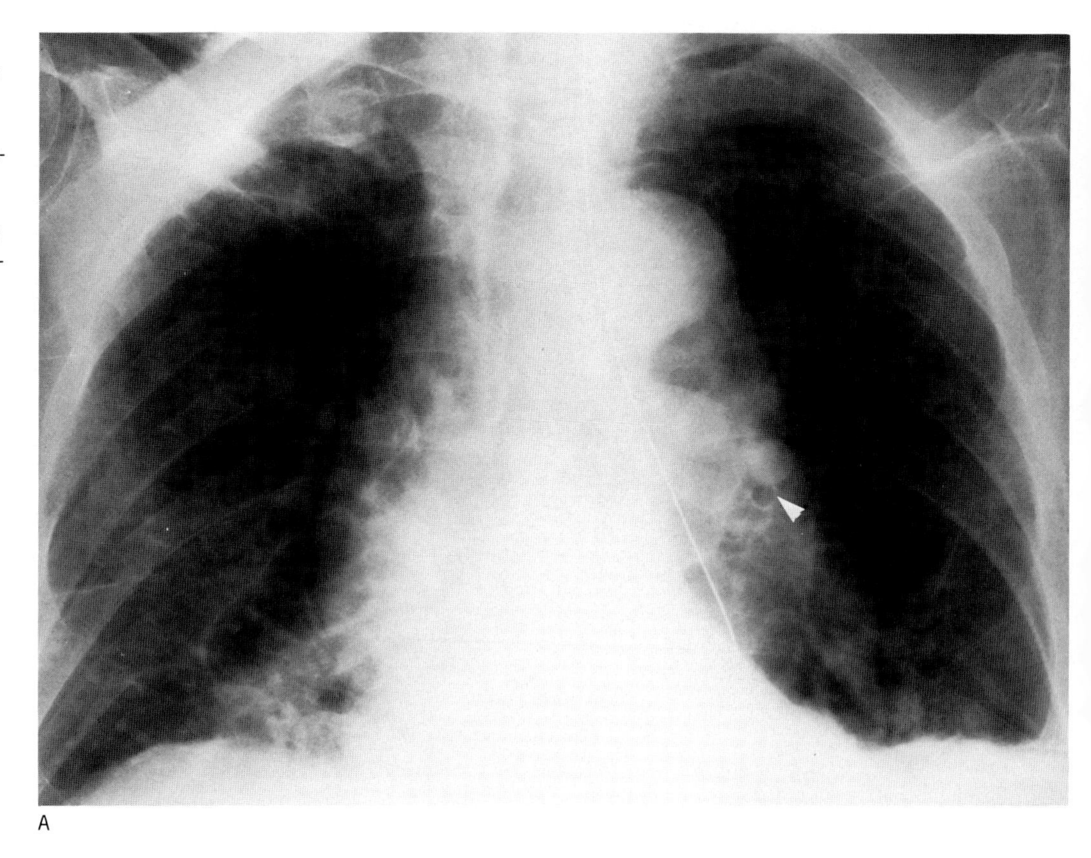

A

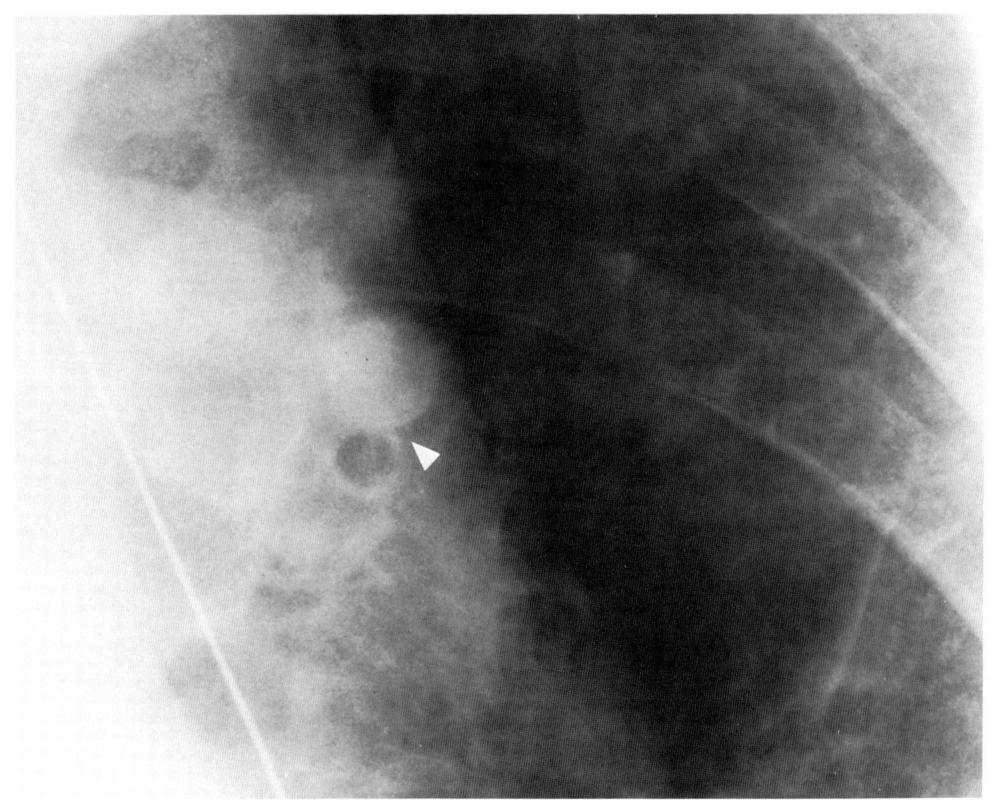

B

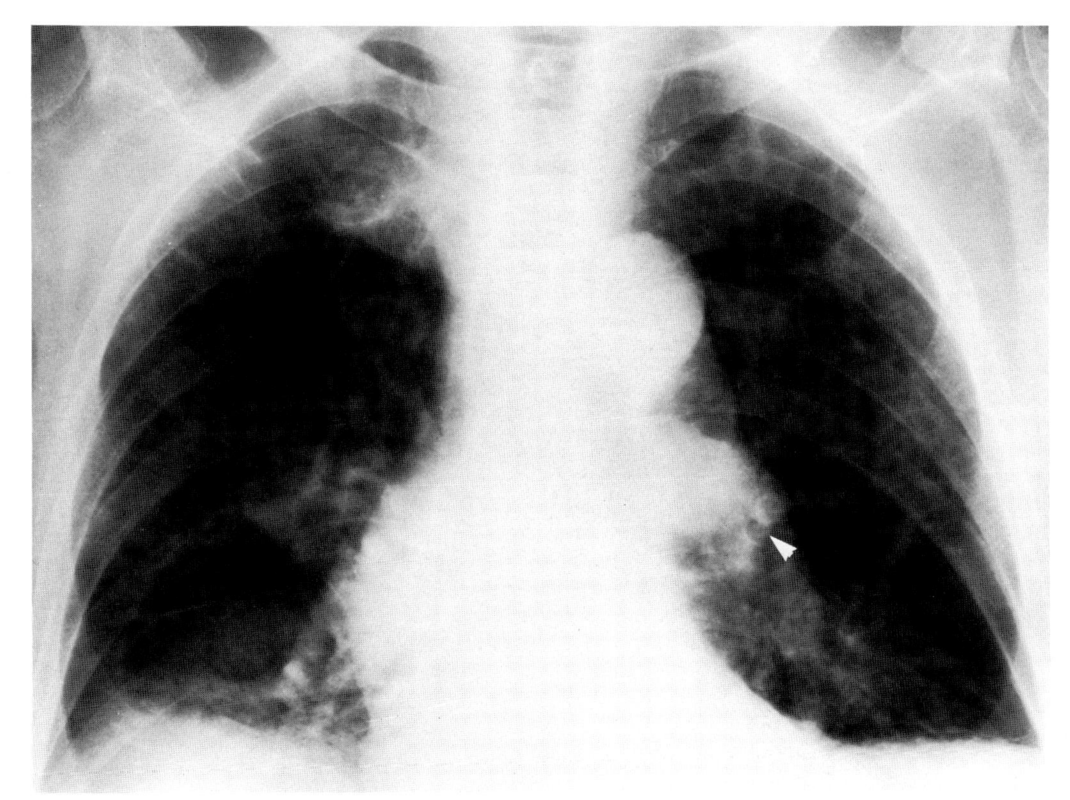

C

One should also compare the caliber of the upper lobe vessels with the lower lobe vessels. The caliber of the upper lobe vessels is smaller than that of the lower lobes by a ratio of 2:3. This fact is useful only when viewing an upright chest film, because of the difference in the flow between areas affected by gravity. With the development of pulmonary venous hypertension, changes become noticeable at pressures of 10 to 13 mm Hg, producing equalization of the size of these vessels [48]. At pressures of 14 to 20 mm Hg, cephalization (diameter of upper lobe vessels exceeds that of the lower lobes) occurs [48] (Fig. 3-13).

Sharpness of the vascular outlines is another gauge of pulmonary venous pressure. At pressures of 20 to 25 mm Hg, fluid exuding into the interspaces begins to blur the vascular outlines [48]. At these pressures, the normal pencil line–thin width of the wall of the bronchus-on-end increases in thickness. This is called peribronchial cuffing (Fig. 3-14). However, peribronchial cuffing is nonspecific and may also be seen with inflammation or fibrosis. Septal interstitial edema (Kerley A, B, and C lines) is also seen in pulmonary venous hypertension (Fig. 3-15). Alveolar edema results when pulmonary venous pressure reaches a critical level of approximately 25 mm Hg (Fig. 3-16).

Fig. 3-13. Vascular redistribution. Arrows point to the upper lobe vessels that are of greater caliber than the vessels (arrowhead) in the right lower lobe.

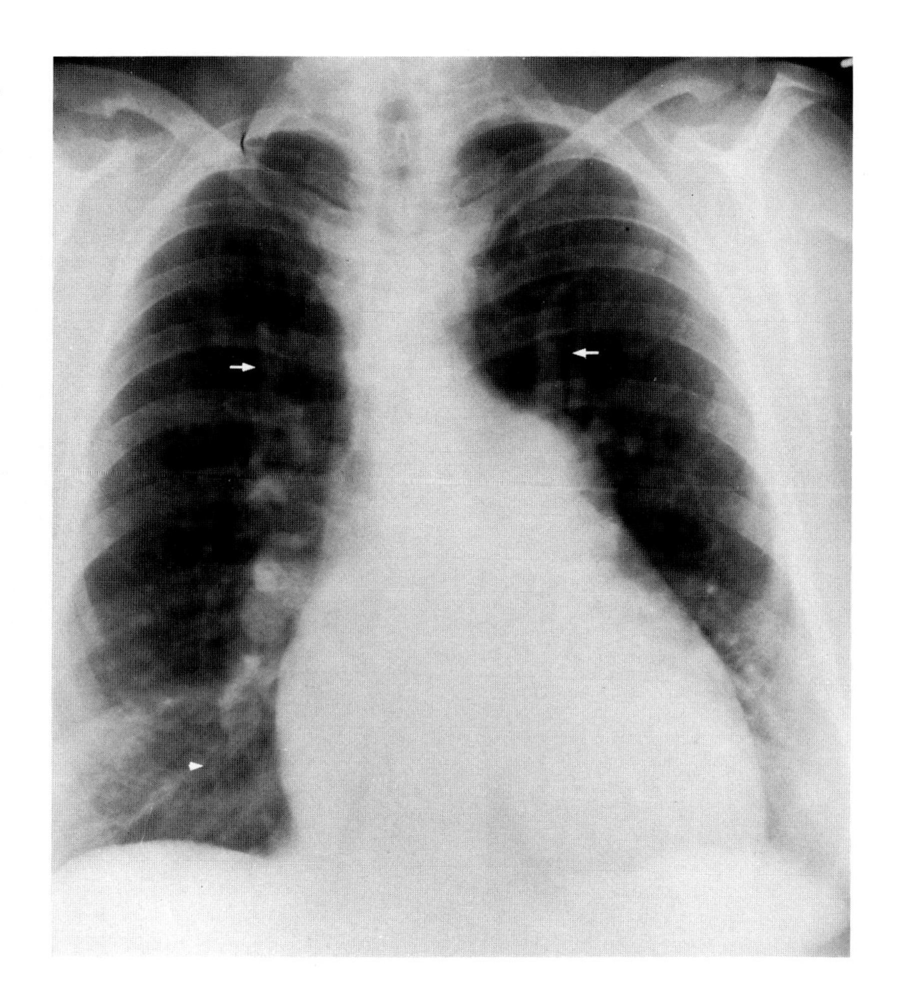

Fig. 3-14. Cuffed bronchus-on-end (arrow) beside a vessel-on-end (arrowhead).

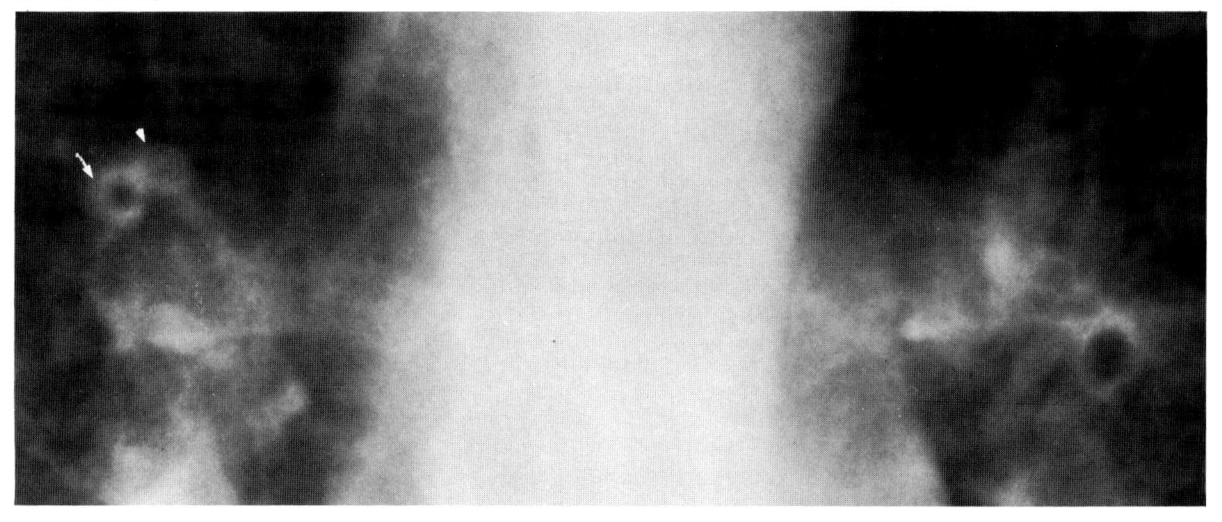

Fig. 3-15. *A.* Patient with interstitial edema. *B.* Close-up of Kerley B lines (arrowheads).

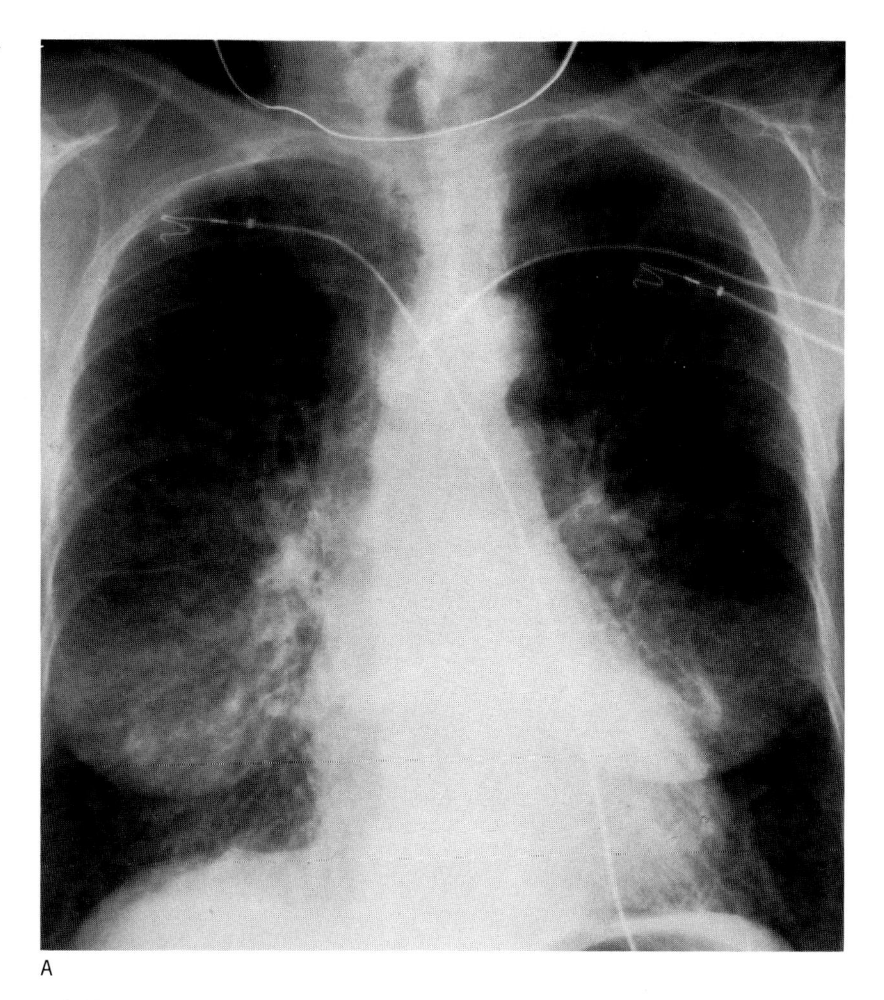

A

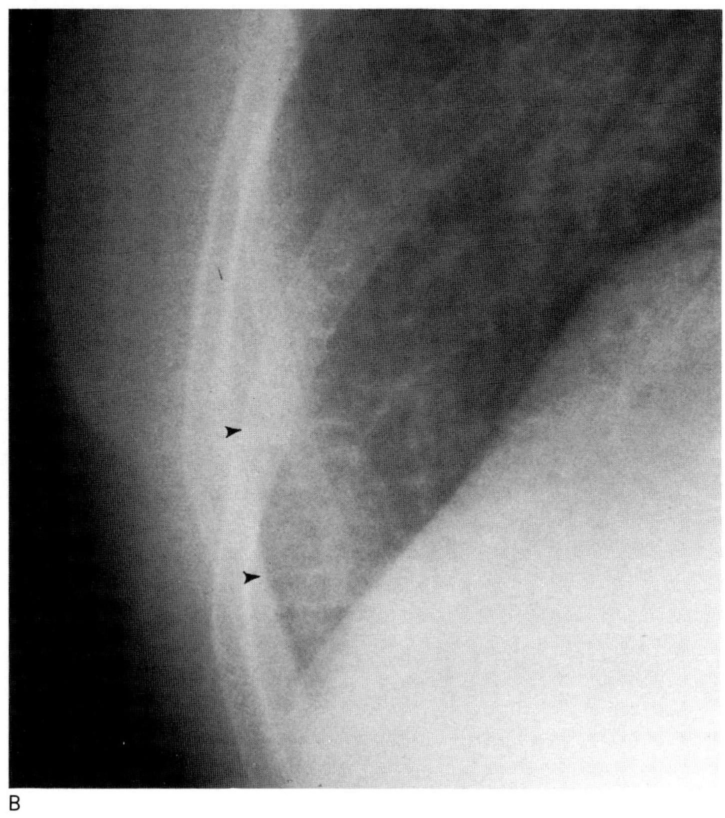

B

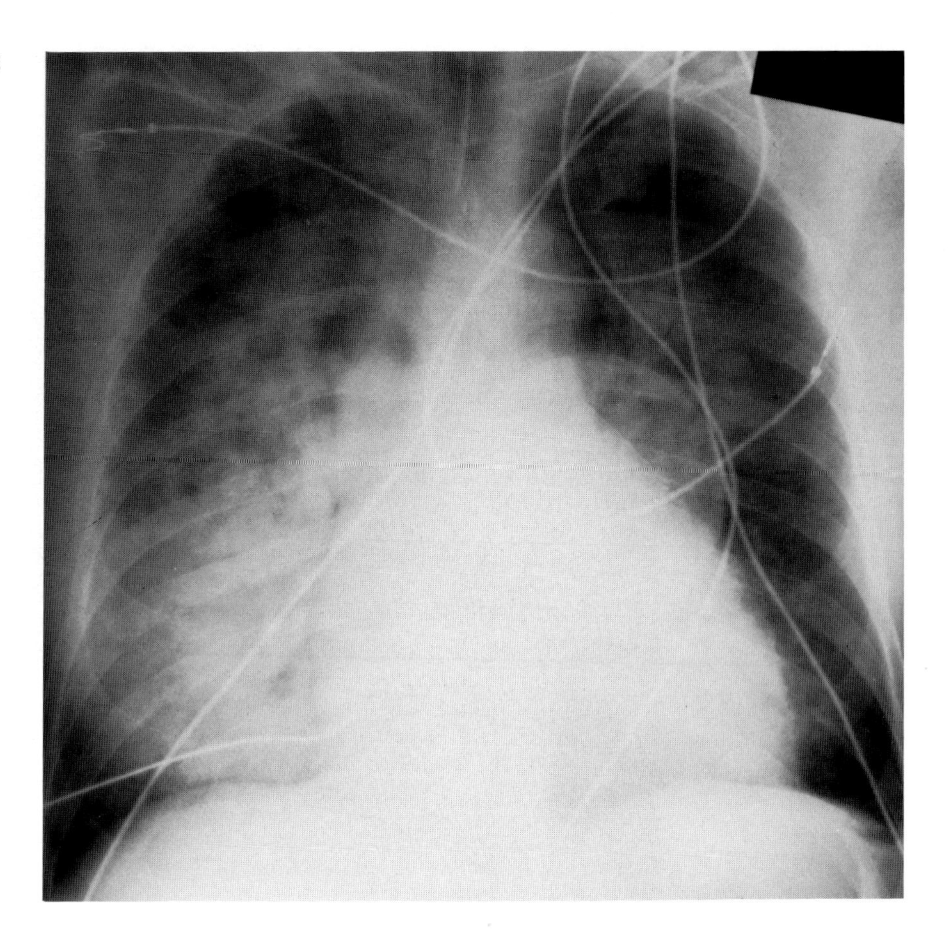

Fig. 3-16. Butterfly or bat-wing pattern of alveolar pulmonary edema.

AZYGOUS VEIN

The width of the azygous vein also correlates with central venous pressure. The azygous vein is located at the point where the right upper lobe bronchus branches off from the trachea (Fig. 3-17). The azygous node is found at the same location on the PA view. To determine which one is enlarged, one can either get upright and supine films or fluoroscope the patient while performring Valsalva's or Müller maneuvers. If the enlarged structure is the azygous vein, it will change in size between the upright and the supine films or in fluoroscopy with the above maneuvers. Also, the azygous node is an anterior structure (Fig. 3-18), whereas the vein is posterior. The normal azygous vein measures 7 mm in the upright position, except in pregnant women, in whom it can reach 15 mm [49]. Preger and associates [50] showed that central venous pressure (CVP) can be calculated within 95 percent confidence limits by measuring the azygous vein. Their formula is as follows:

$$(\text{Azygous vein width in mm} \times 1.4) - 3 = \text{CVP cm } H_2O.$$

Fig. 3-17. Normal azygous vein (arrowheads).

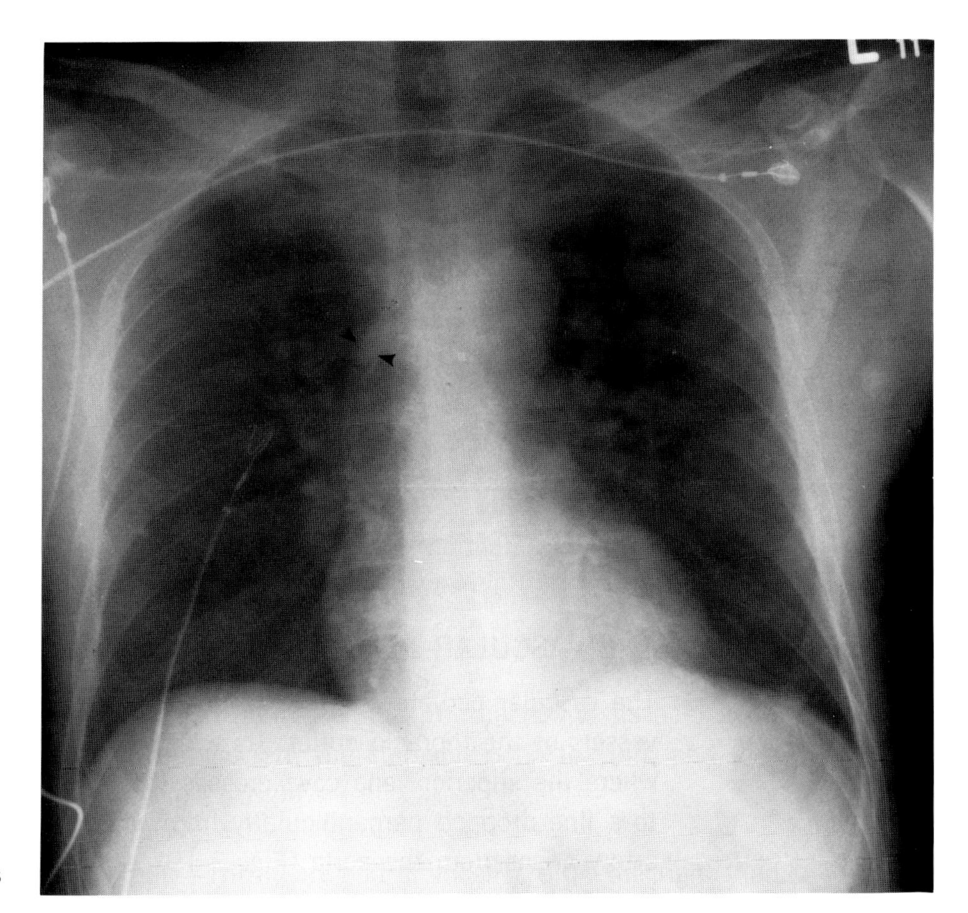

Fig. 3-18. *A*. The azygous node is located in the same area as the azygous vein and is almost indistinguishable from the vein except for a more anterior position in the lateral view (*B*). Arrows point to a calcified azygous node. *B*. Lateral view of the patient in *A*.

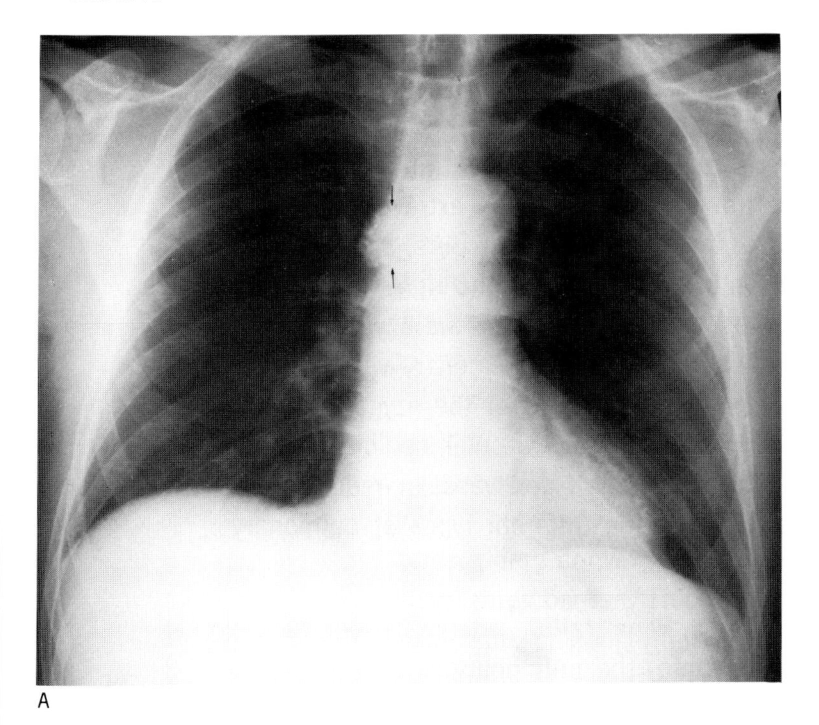

A

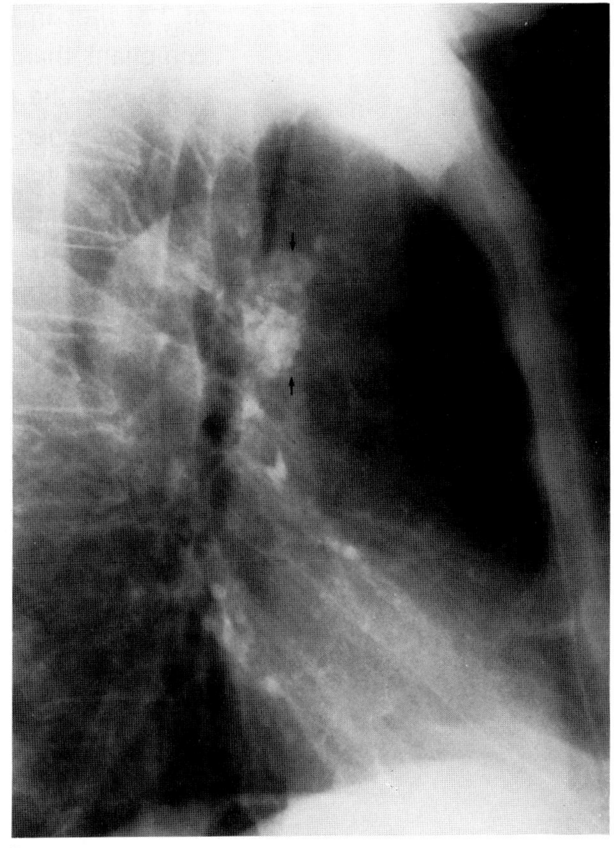

B

CARDIAC CHAMBERS

The size of cardiac chambers should be individually assessed.

Left Ventricle Enlargement The left ventricle is the most frequently enlarged chamber in the adult heart. When it enlarges, it does so posteriorly and laterally. The most useful sign for left ventricular enlargement was described by Hoffman and Rigler [55]. It consists of two measurements made in the lateral view. Measurement A is defined as the distance between the left ventricle posteriorly and a point 2 cm directly cephalad from the crossing of the inferior vena cava and the left ventricle (Fig. 3-22). If A exceeds 1.8 cm, then the left ventricle is considered enlarged (Fig. 3-23A). Measurement B is the distance from the crossing of the inferior vena cava to left ventricle caudally (Fig. 3-23B). If this distance is less than 0.75 cm, left ventricular enlargement may have occurred. Change in contour of the cardiac border also indicates abnormality (Fig. 3-24, see also Fig. 4-10).

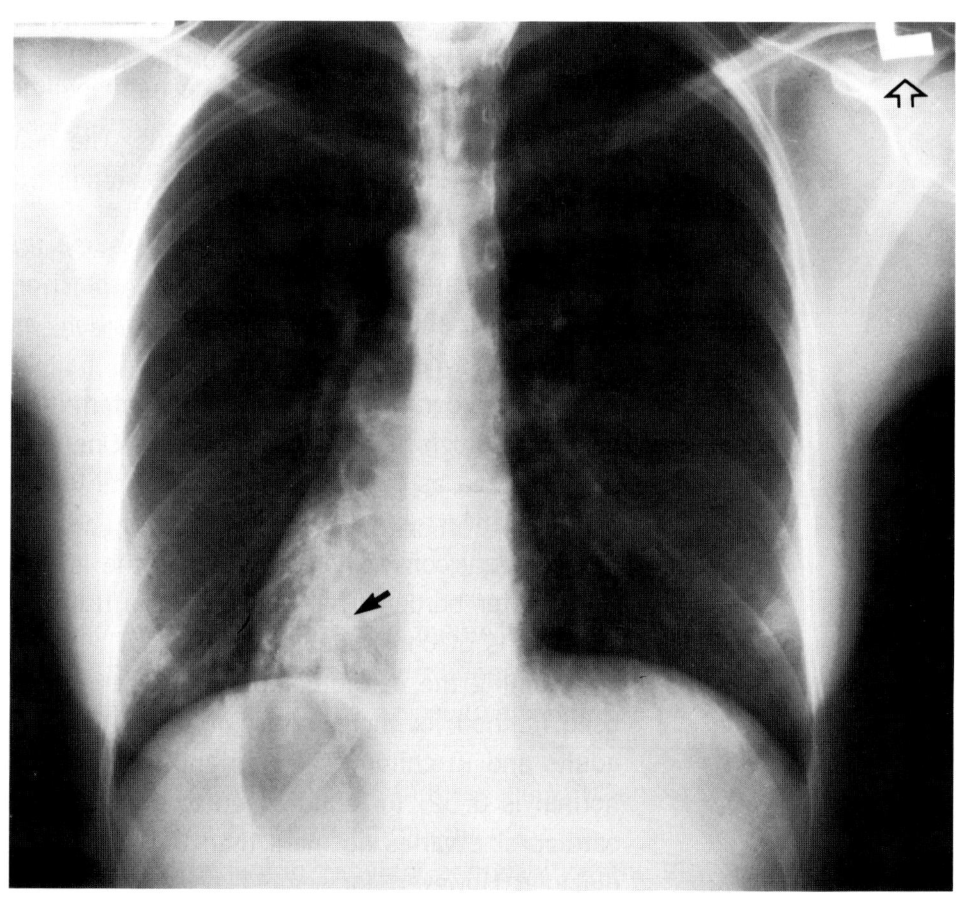

A

Fig. 3-20. *A.* Patient with Kartagener's syndrome. Open arrow points to left marker. Note situs inversus. Arrow points to the right lower lobe bronchiectasis. *B.* Patient with situs indeterminatus. Note symmetrical liver shadow and bilateral right-sidedness of the lungs.

Left Atrial Enlargement When the left atrium enlarges with the concurrent dilatation of the pulmonary veins, the left bronchial tree is often displaced posteriorly on the lateral view (Fig. 3-25). Another method of determining left atrial enlargement is to use the right pulmonary artery/left atrial axis line. If the anterior wall of the right pulmonary artery on the lateral view can be identified clearly, one can draw a line from its anterior margin parallel to the lower third of the esophagus. Another line, perpendicular to the first, can be drawn to the posterior wall of the left atrium. In infants and

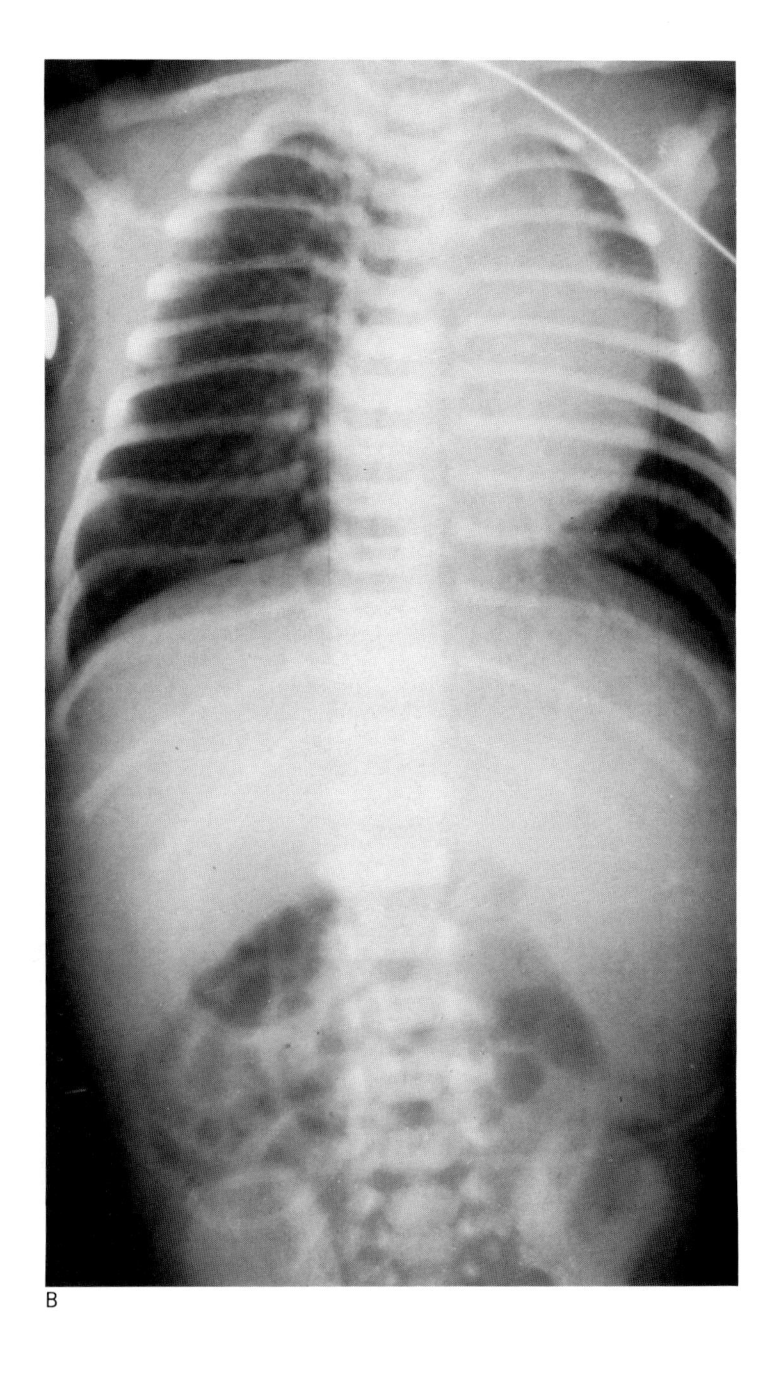

B

Fig. 3-21. *A.* Line 1 is the transverse diameter of the heart and line 2 is the widest transverse diameter of the chest. CT ratio is obtained by dividing the length of line 1 by the length of line 2. In this case the ratio is abnormal. *B.* Line A is through the midline; line B to the farthest right cardiac border; and line C to the farthest left cardiac border. B and C = transverse cardiac diameter; D = widest transverse diameter of the chest. B + C/D = cardiothoracic ratio.

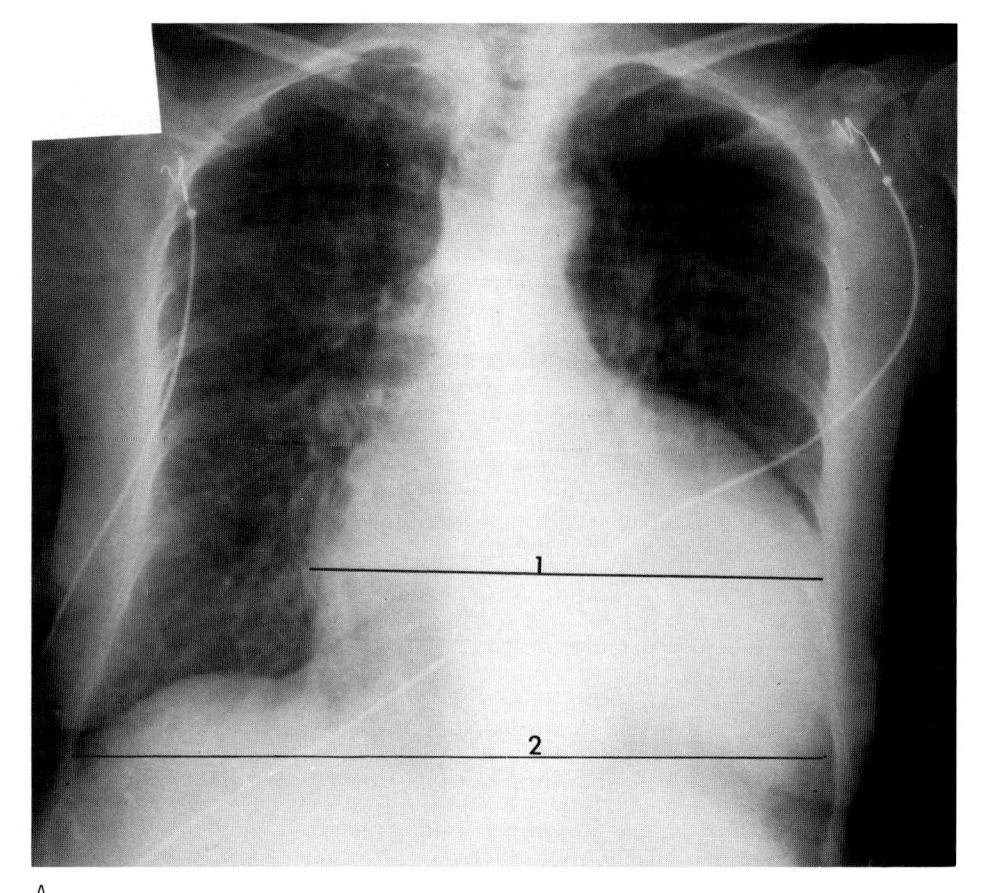

A

B

Fig. 3-22. *A.* Normal lateral view of chest. Arrow points to area where inferior vena cava meets the cardiac silhouette. *B.* Close-up of the lower, posterior cardiac border in the lateral view. 1 traces the outline of the diaphragm; 2, the outline of the inferior vena cava; and 3, the outline of the posterior cardiac silhouette. Arrow points to where the cardiac silhouette and inferior vena cava meet.

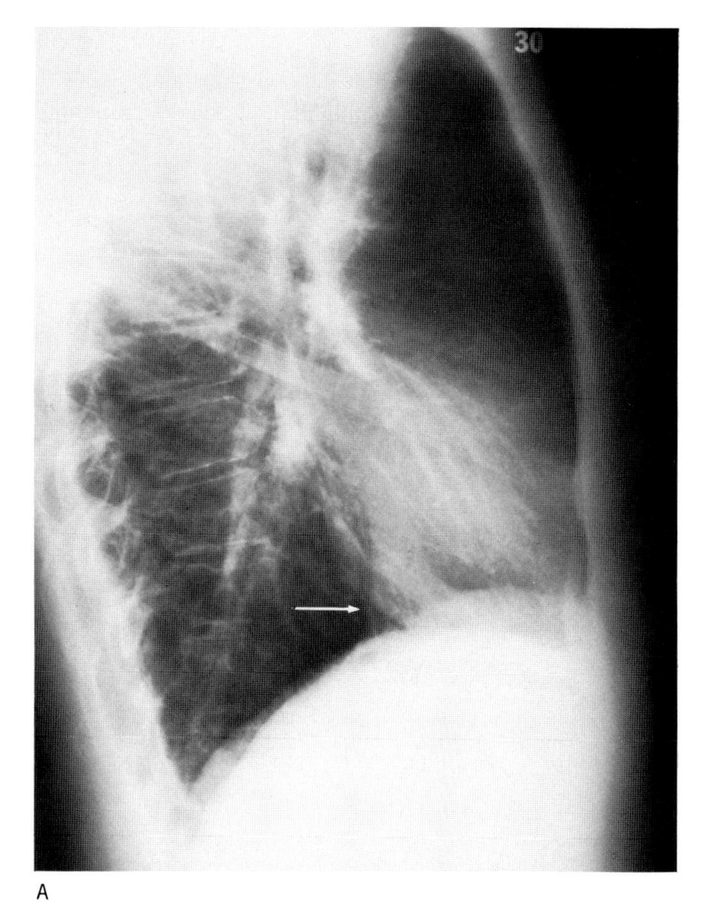

A

B

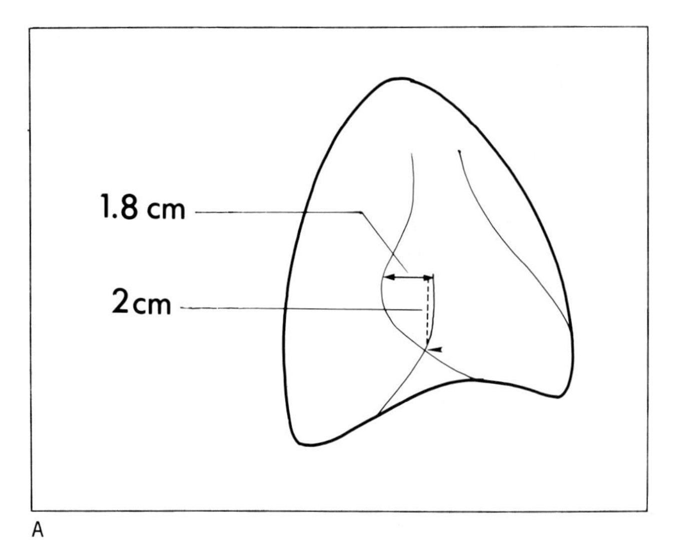

A

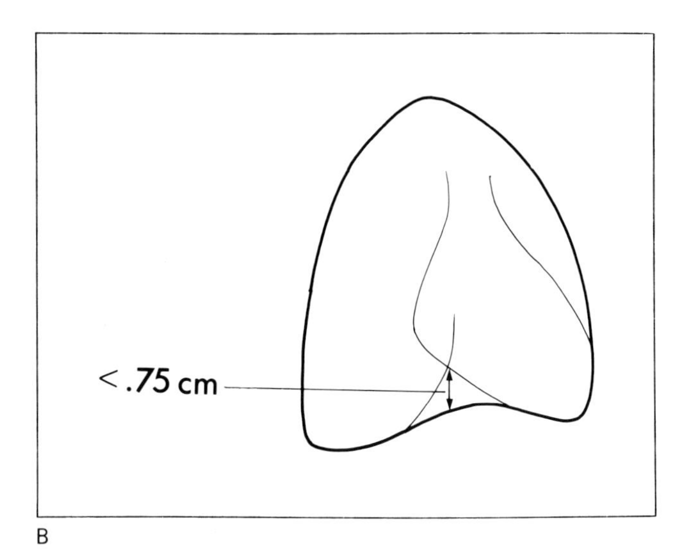

B

Fig. 3-23. *A.* Assessment of left ventricular enlargement. Measurement A: Arrowhead indicates crossing of inferior vena cava and left ventricle. A 2 cm line is drawn cephalad from this point (broken line). A normal-sized left ventricle does not exceed 1.8 cm posteriorly from this point. (Adapted from Hoffman and Rigler [55].) *B.* Measurement B is the distance from the crossing of the inferior vena cava and left ventricle (upper arrow) to the diaphragm (lower arrow). If this distance is less than 0.75 cm, then it is additional proof of left ventricular enlargement.

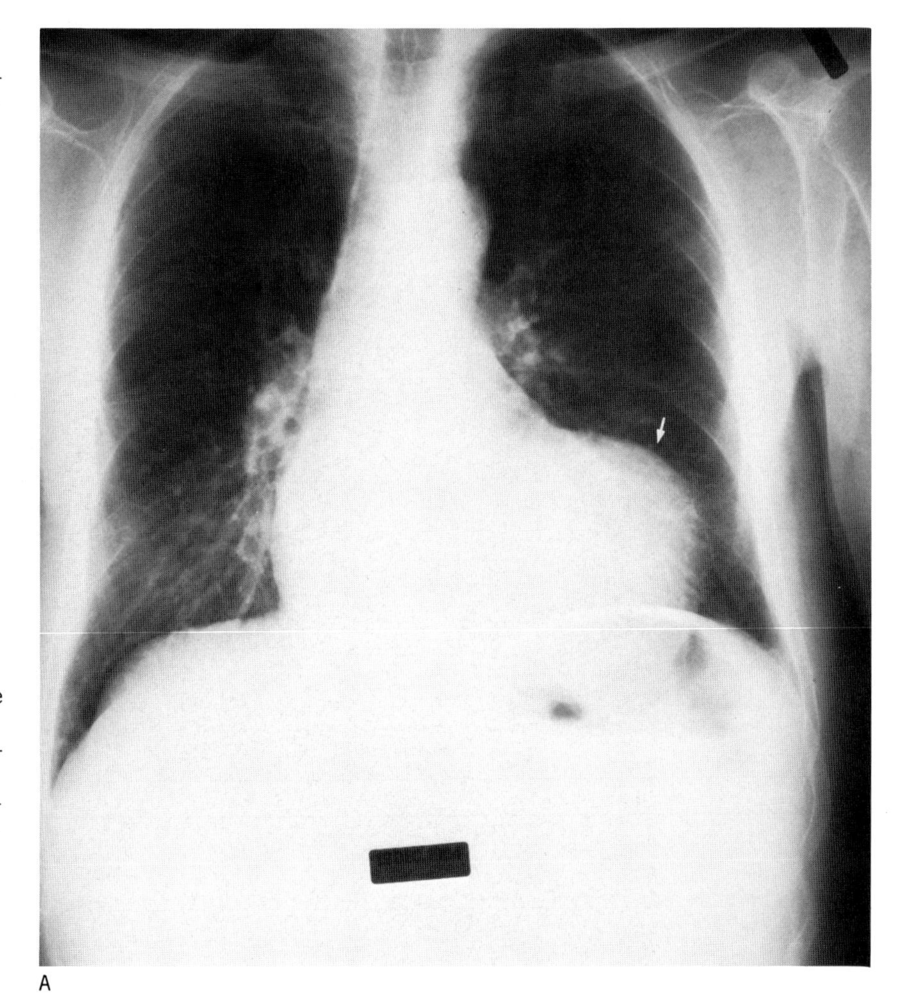

A

Fig. 3-24. *A.* PA view shows an abnormal bulge (arrow) in the outline of the left ventricle secondary to a left ventricular aneurysm. *B.* Lateral view. Arrow points to the double density produced by the left ventricular aneurysm. *C.* Another patient with a left ventricular aneurysm. Arrow points to a protusion in the left cardiac border from a left ventricular aneurysm.

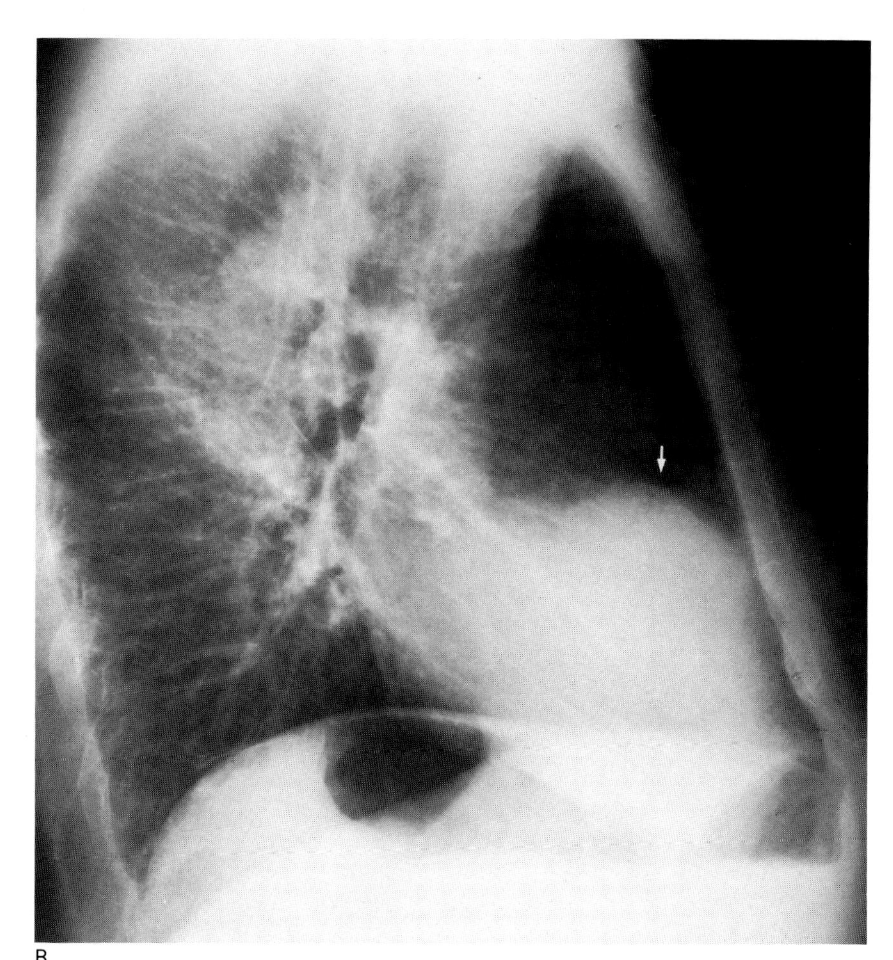

B

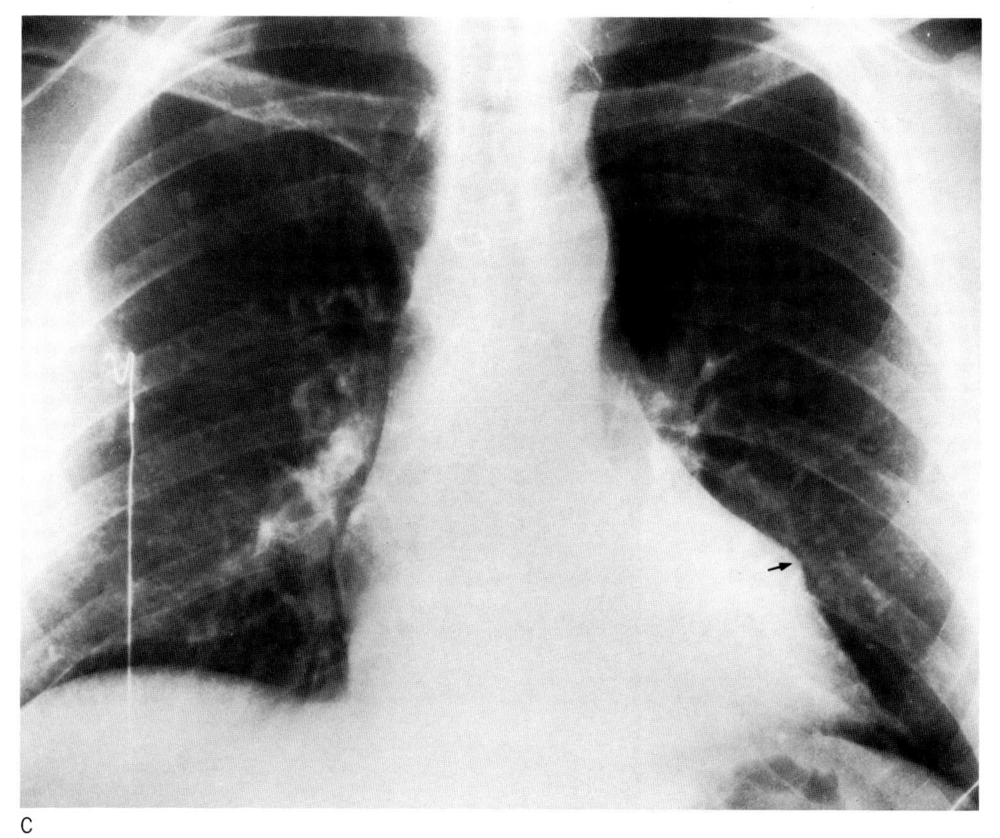

C

Fig. 3-25. Posteriorly displaced left bronchus (arrowhead) and enlarged left atrium (arrow). Tip of arrow points to the posterior extent of the left atrium.

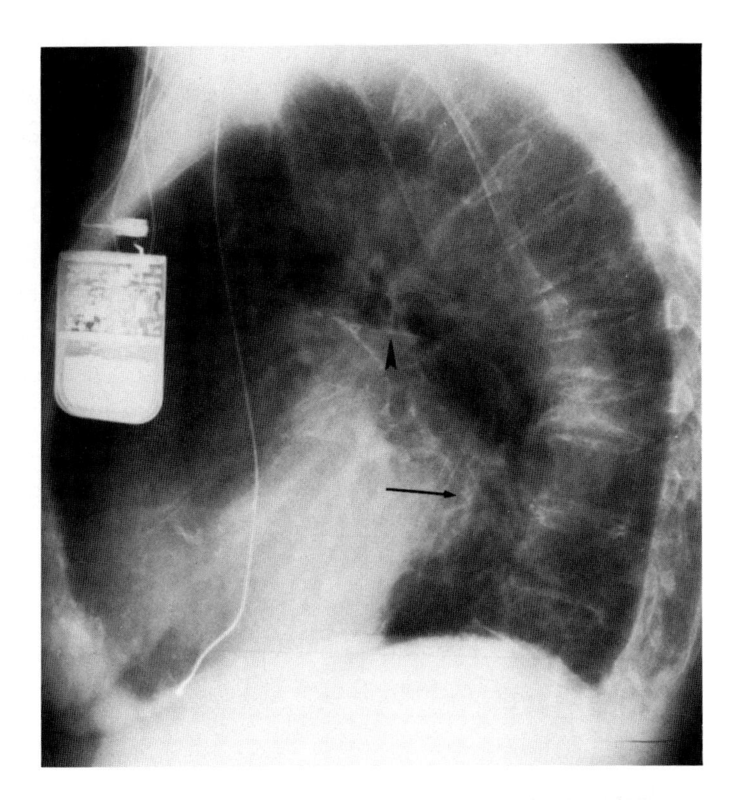

Fig. 3-26. Line 1 is from the anterior margin of the right pulmonary artery, parallel to the course of the esophagus. 2 represents line perpendicular to 1, drawn to the posterior wall of the left atrium.

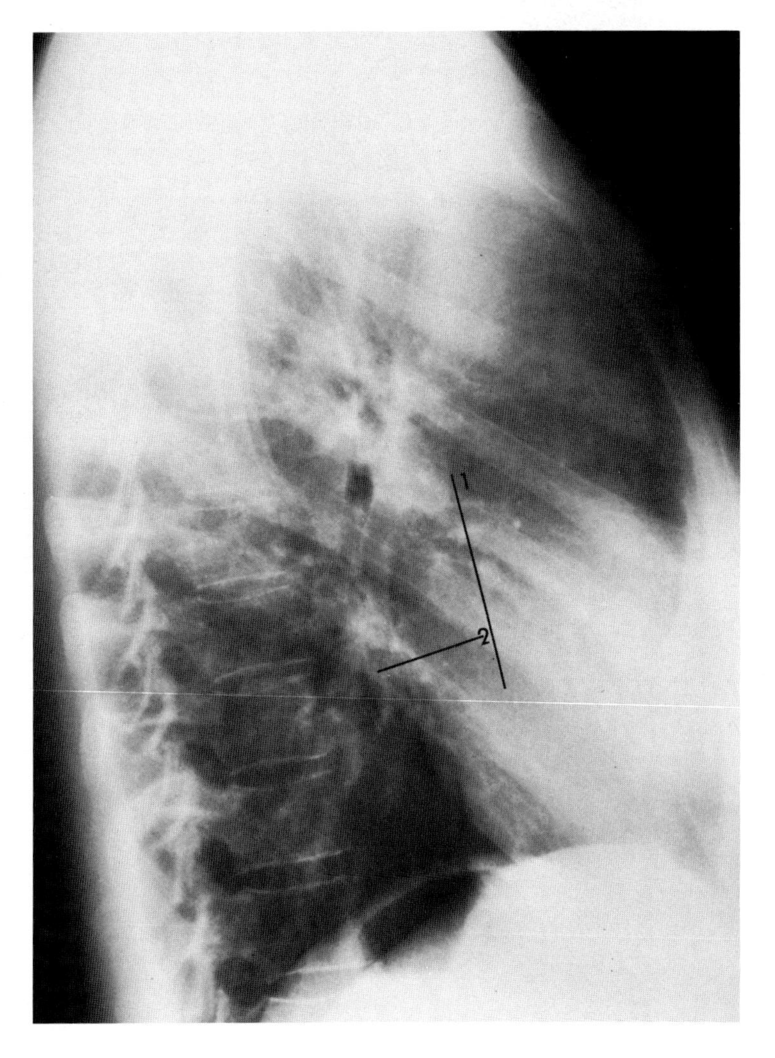

children, the normal range for the distance between these two lines is 2 to 3 cm; in adults, the maximum value is 3.8 cm for females and 4.5 cm for males [56] (Fig. 3-26).

Left atrial enlargement is seen in the PA view as a double density (Fig. 3-27) (in some children, the confluence of pulmonary veins may produce a double density despite a normal left atrium). The carinal angle may increase in some. Measuring the distance from the right wall of the double density to the left main stem bronchus in the PA view can also aid in evaluating left atrial enlargement. The upper value of normal is 3.5 cm in infants, 4.5 cm in children, and 7.5 cm in adults [57]. Values greater than these are considered abnormal.

Right-side Enlargement It is difficult to diagnose isolated right-side enlargement. There is also little need to separate the right atrium from the right ventricle, because few lesions

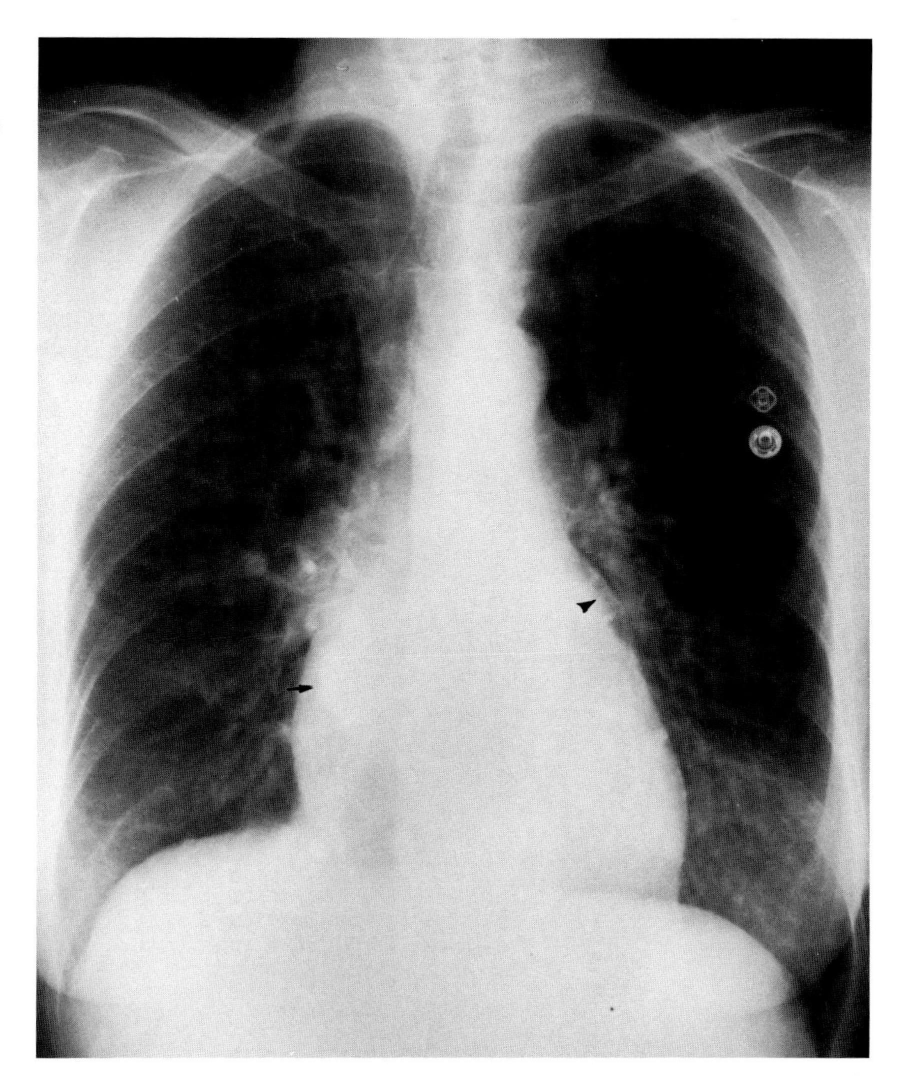

Fig. 3-27. Arrow points to the double density produced by the enlarged left atrium; arrowhead to the convexity produced by the left atrial appendage.

Fig. 3-28. *A*. PA view; arrow points to the increased convexity of the right cardiac border in right ventricular enlargement. *B*. Lateral view shows the cardiac silhouette occupying 50 percent of the retrosternal space.

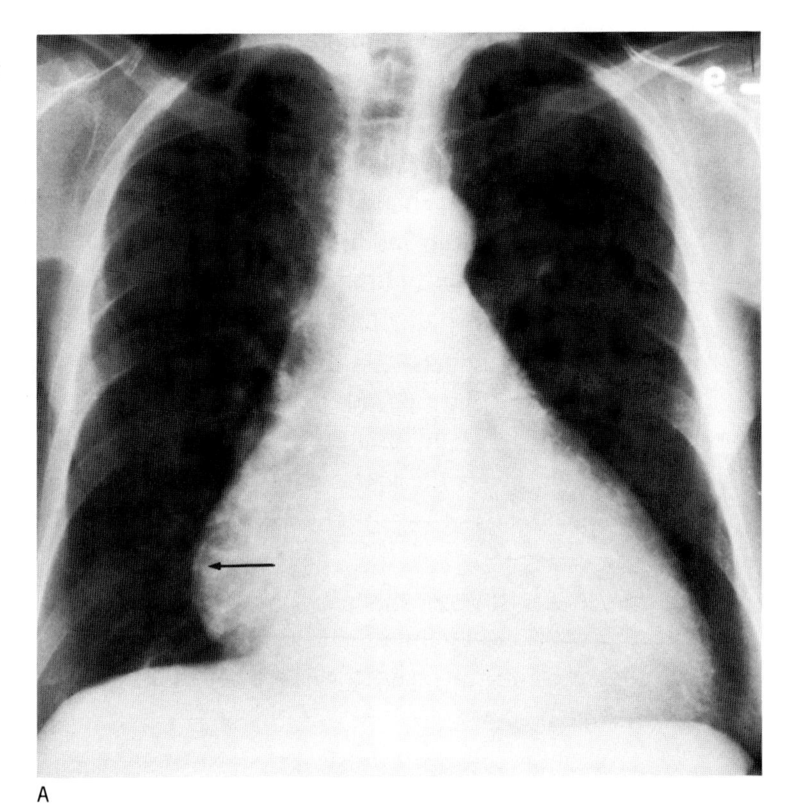

A

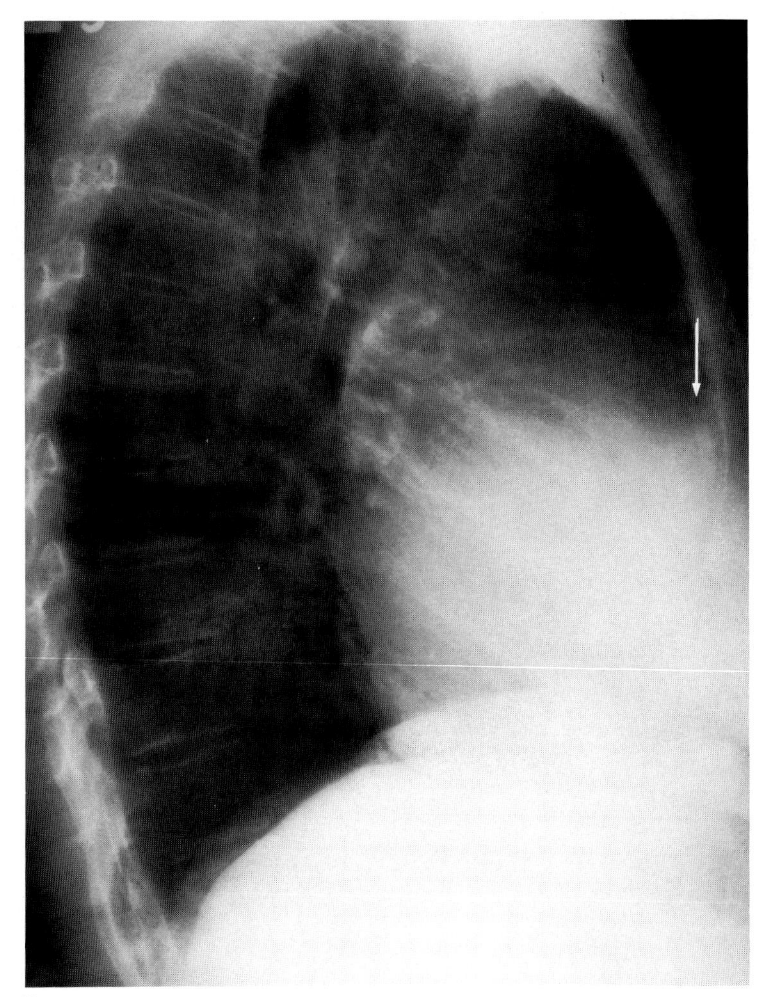

B

yield isolated right atrial enlargement (Ebstein's malformation). The term *right-side enlargement* will, then, encompass both right atrial and right ventricular enlargement.

In the PA view, one sees increased convexity of the right cardiac border. The right ventricle becomes border forming, and normally the indentation between the aorta and the pulmonary artery is obliterated (Fig. 3-28). In the lateral view, the right ventricle normally occupies 30 to 40 percent of the retrosternal space (the distance from the manubriosternal junction to the anterior diaphragmatic surface). When it exceeds 40 percent, one considers right-side enlargement. However, this sign of right ventricular enlargement (Fig. 3-28B) cannot be used for evaluating infants and children, because the normal thymus tissue in young people occupies the anterior mediastinum and silhouettes out the right cardiac border.

The Pericardium The normal pericardium is often well visualized in the lateral projection, and occasionally in the PA projection, as a thin strip of density, 1 to 2 mm thick, sandwiched between the lucent epicardial fat and the mediastinal fat [58] (Fig. 3-29). An increase in the thickness of the density to greater than 2 mm suggests pericardial effusion or

Fig. 3-29. *A.* Normal pericardial thickness. *B.* PA view; close-up of area of pericardium. Epicardial fat (1), pericardium (2), mediastinal fat (3).

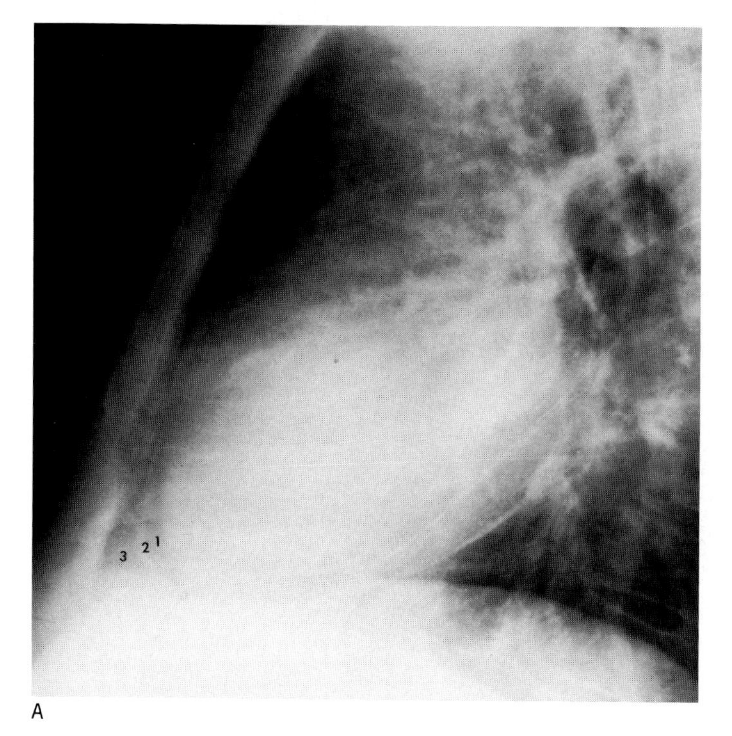

A

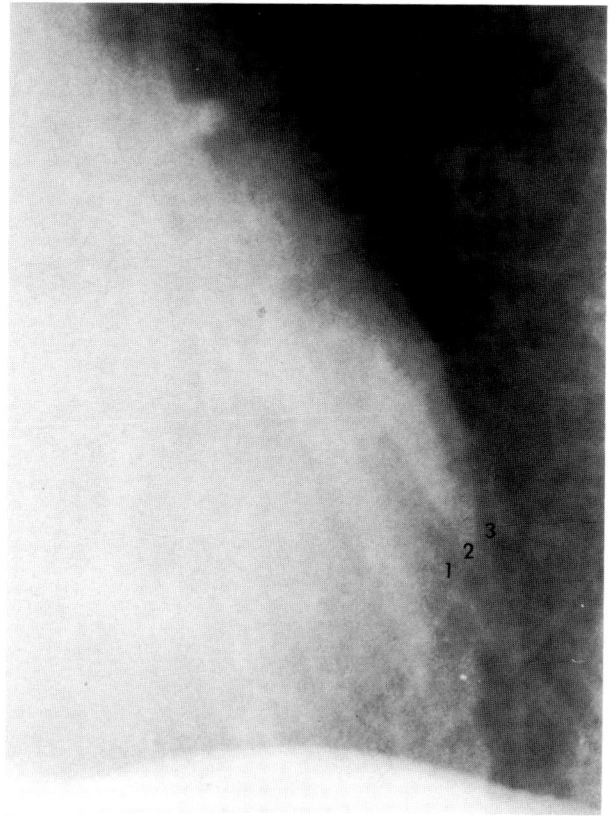

B

Fig. 3-30. *A.* A normal pericardial shadow on the lateral view (arrow). *B.* Arrowheads at the edges of the thickened pericardium (arrow) in the same patient with pericardial effusion.

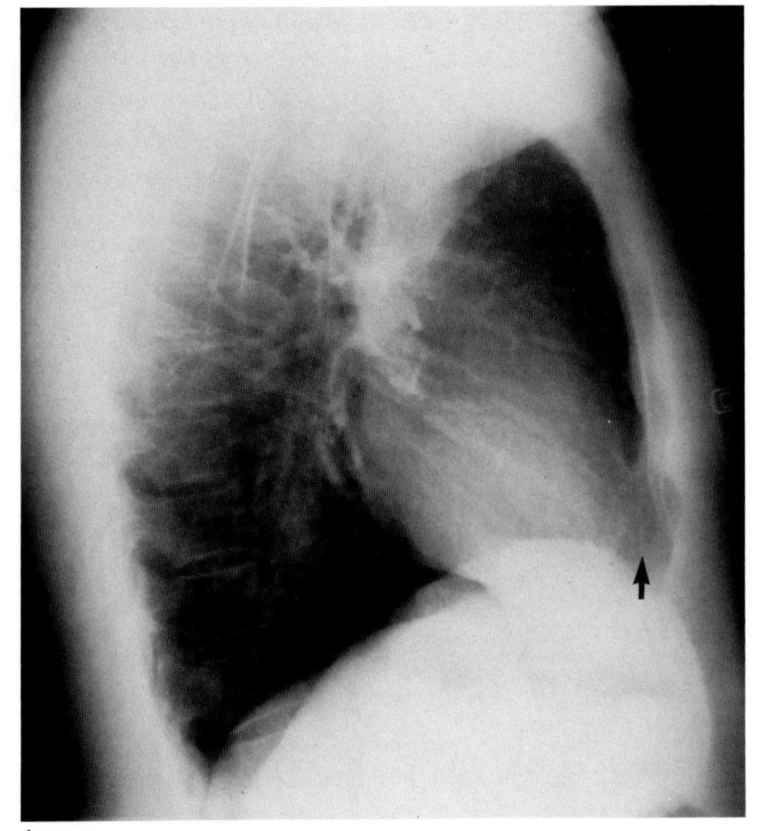

A

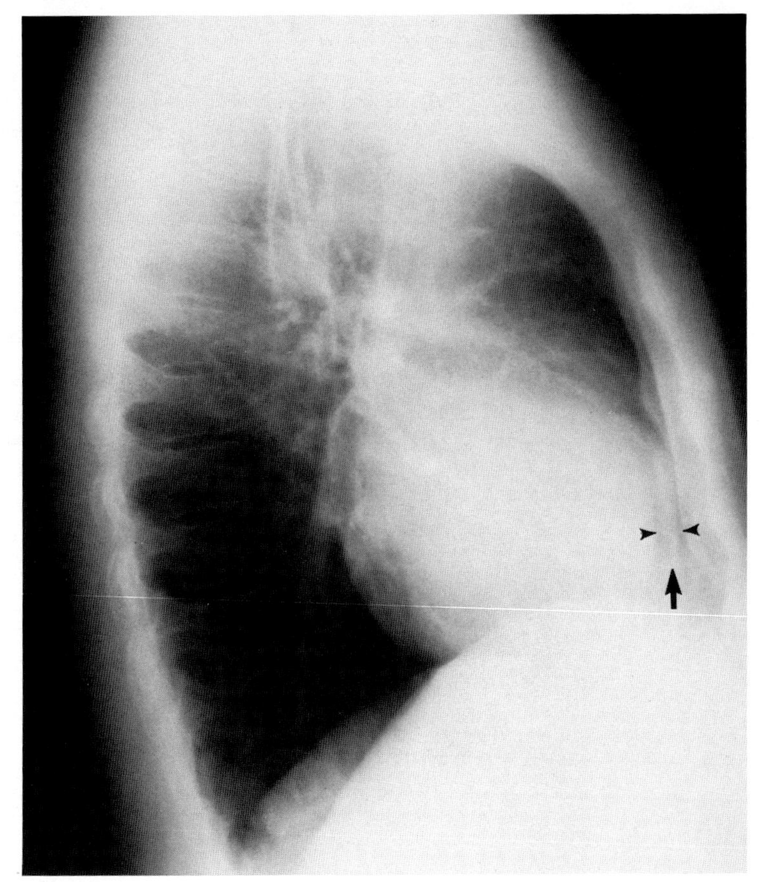

B

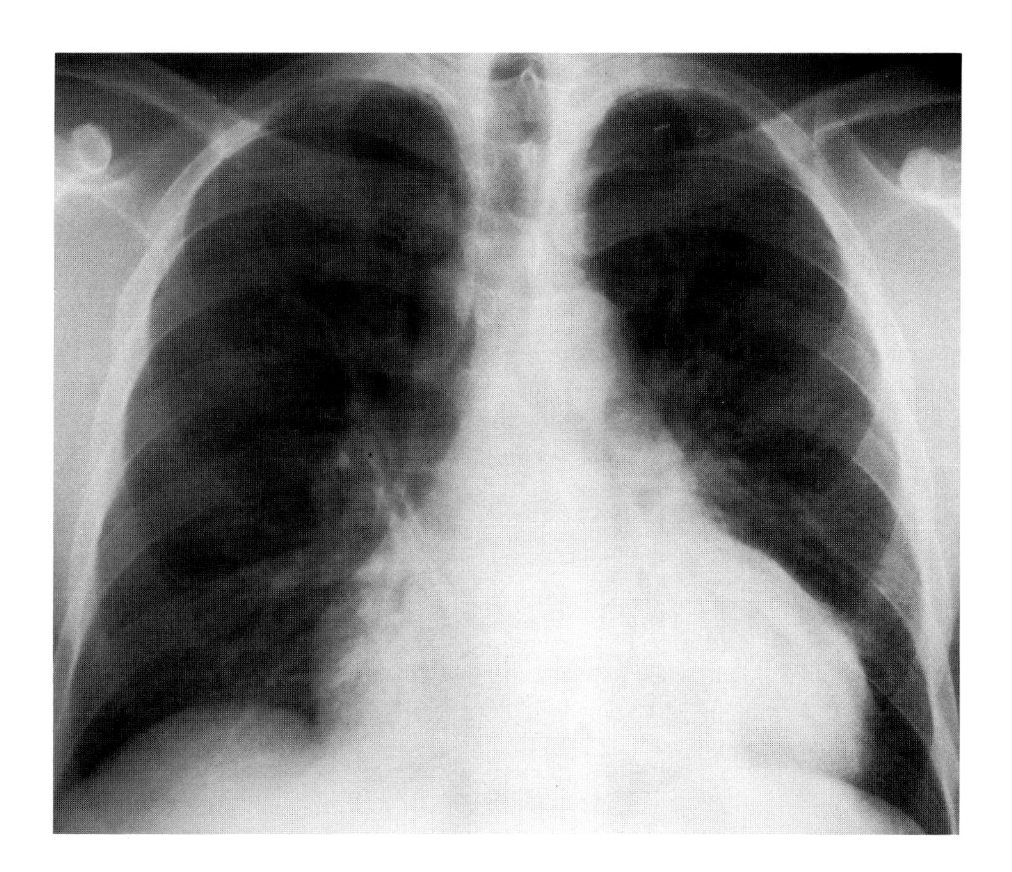

Fig. 3-31. Water-bottle appearance of the cardiac silhouette with pericardial effusion.

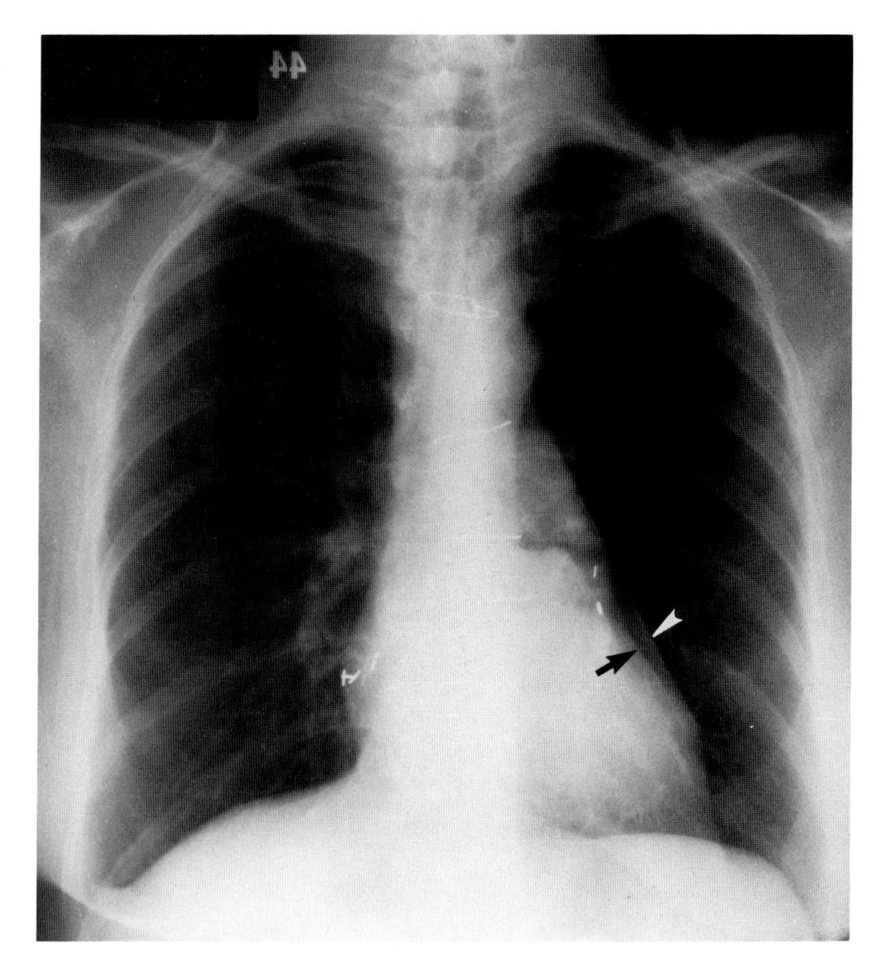

Fig. 3-32. Arrowhead points to the pericardium; arrow to the lucent air between the heart and the pericardium. Pneumopericardium is secondary to coronary artery bypass surgery.

thickening [59] (Fig. 3-30). A typical water-bottle configuration in the PA view is also seen in pericardial effusion (Fig. 3-31).

The radiologic diagnosis of a pneumopericardium is made when a lucent stripe is seen around the heart, extending to, but not beyond, the proximal pulmonary artery and outlining a thickened pericardium (Fig. 3-32). It may be difficult to differentiate from a pneumothorax or a pneumomediastinum, and a cross-table lateral film may be necessary to do so. Pneumopericardium is almost always the result of surgery, but may also follow trauma or infection.

STEP 4

EVALUATE THE TRACHEA AND THE MEDIASTINUM

THE TRACHEA

The tracheal air shadow is usually well seen in the PA and lateral views. The inferior border of the glottis forms an arch (Fig. 4-1). It then continues inferiorly as a straight tubular structure (except where indented by the aortic arch) (Fig. 4-2) until it reaches the carina, where it bifurcates into a short right main bronchus and a longer left main bronchus.

Displacement of the tracheal air column may be produced by the following:

1. Normal buckling of the trachea at the thoracic inlet with expiration and neck flexion.

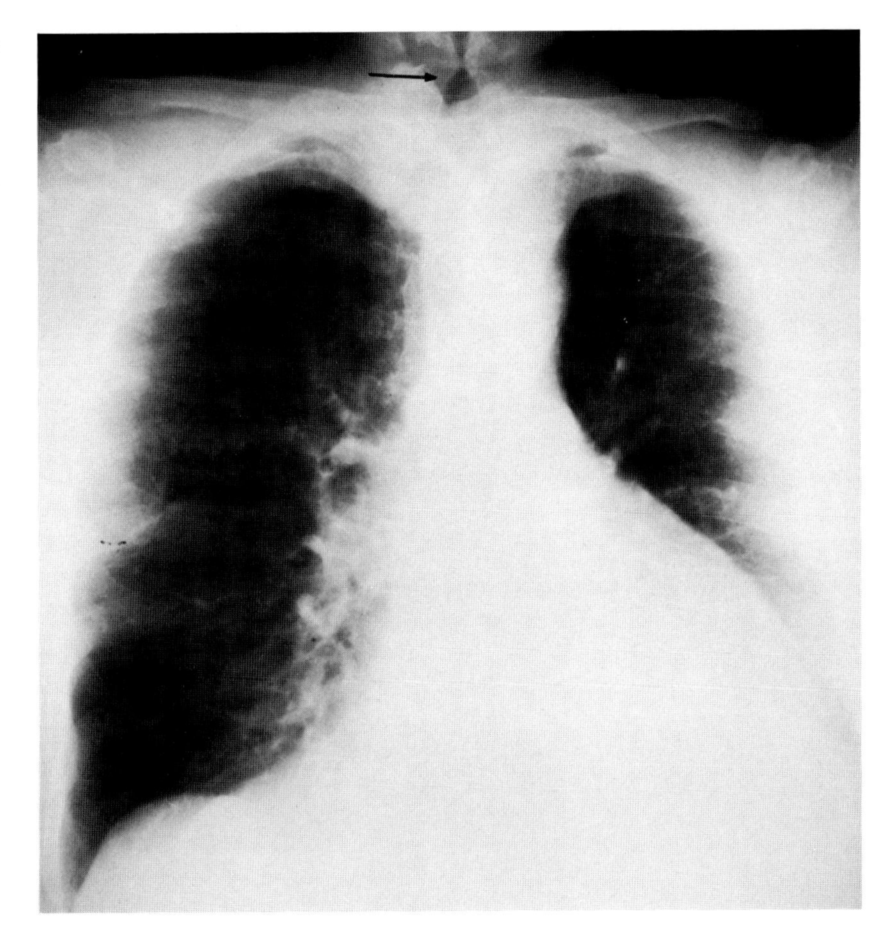

Fig. 4-1. The normal arch of the trachea in the subglottic area (arrow).

Fig. 4-2. *A*. Normal deviation of trachea produced by the aortic arch. *B*. Wide trachea in a patient with Mounier-Kuhn disease (arrowheads). *C*. Sacculations (arrowheads) in the posterior tracheal wall of patient in *B*. *D*. Close-up of the lower lobes demonstrating marked bronchiectasis in same patient.

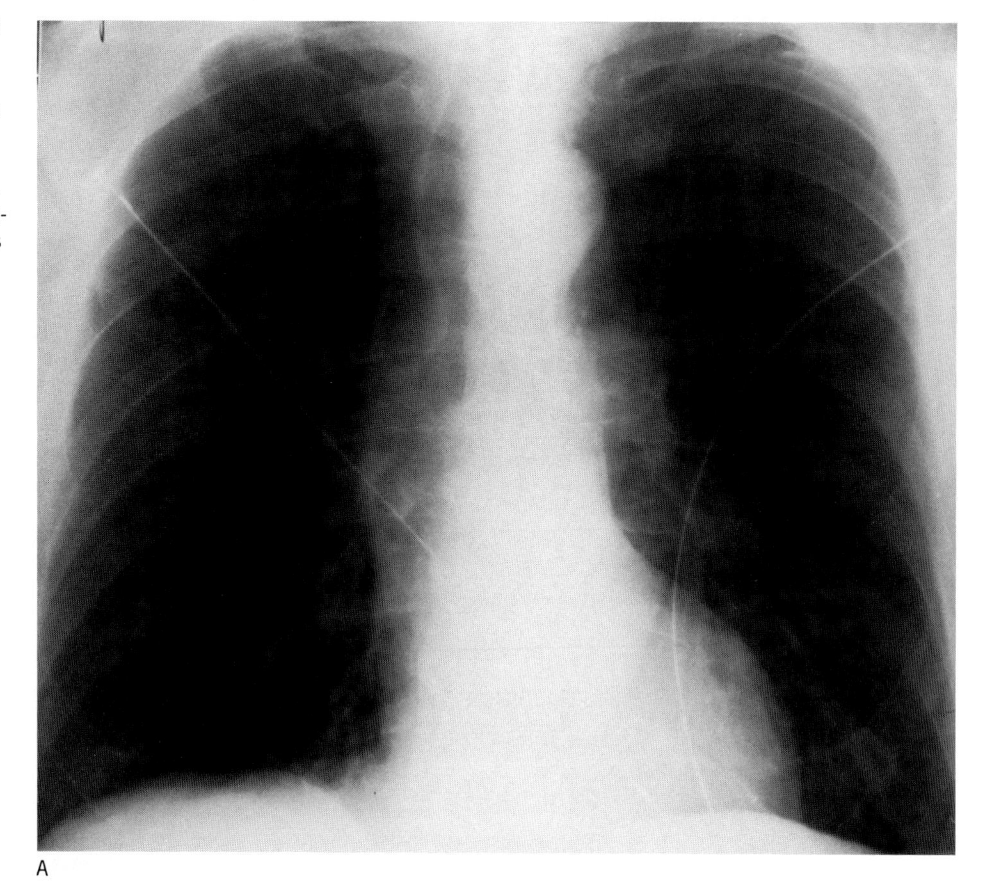

A

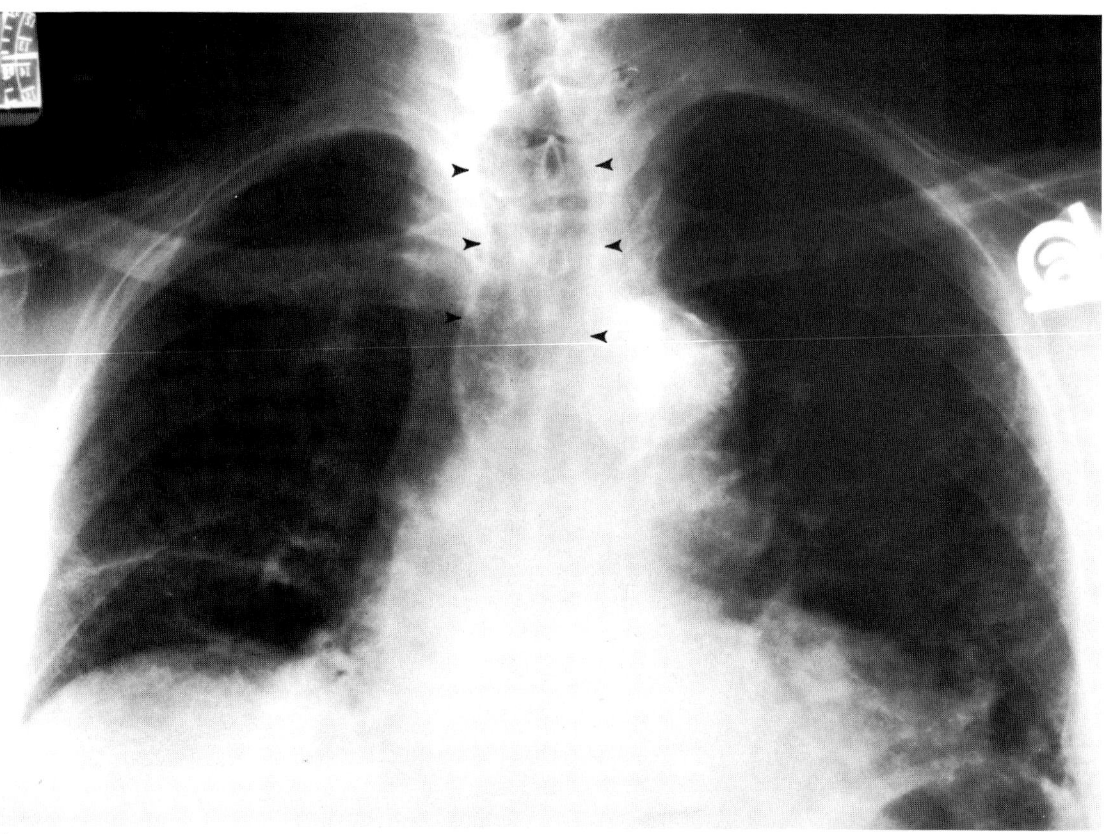

B

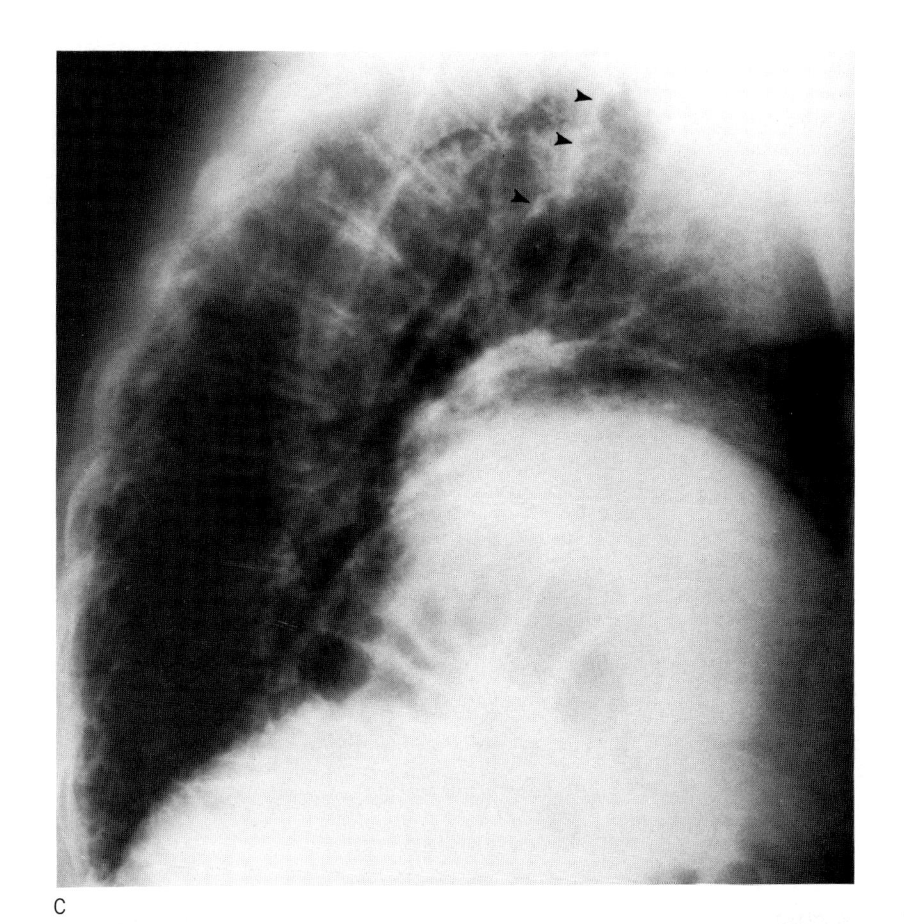

C

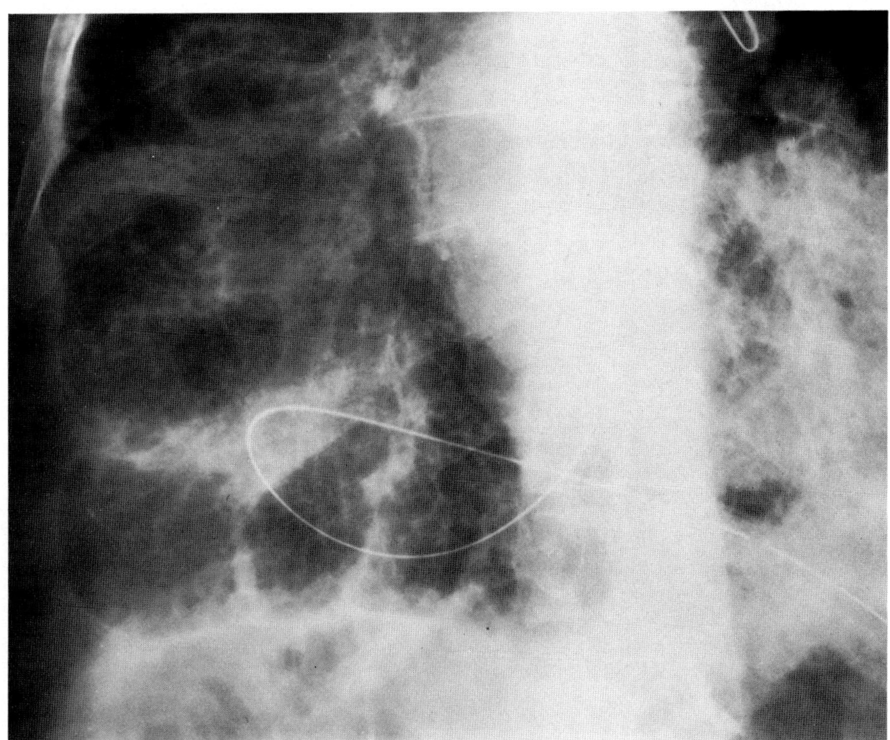

D

2. Mediastinal shifts resulting from changes in intrathoracic volume or pressure or both (e.g., moderate to large areas of atelectasis, lung resection, pleural effusion, pneumothorax).
3. Traction on the trachea caused by inflammatory processes involving nodes, or by fibrosis.
4. Paratracheal masses (e.g., distended pouch in esophageal atresia, duplication cyst of the esophagus, bronchogenic cyst, intrathoracic goiter, and other mediastinal masses adjacent to, or extending into, the region of the trachea).

In children, the normal tracheal diameter varies with age; in adults, it is less than 3 cm.

Widening of the trachea to a diameter greater than 3 cm is seen in Mounier-Kuhn, Ehlers-Danlos, and cutis laxa syndromes, and occasionally in relapsing polychondritis. The transverse or coronal diameter of the trachea alone widens in saber sheath trachea.

Densities seen within the tracheal lumen (Fig. 4-3) may represent the following:

1. Thick, inspissated mucus.
2. Aspirated foreign body.
3. Neoplasm, benign or malignant.

Papilloma, a benign neoplasm with malignant characteristics of spread, is the most frequent pediatric tracheal tumor. Squamous cell carcinomas, bronchial adenomas, and metastatic lesions comprise most lesions in adults. Narrowing of the tracheal lumen may be seen in the following:

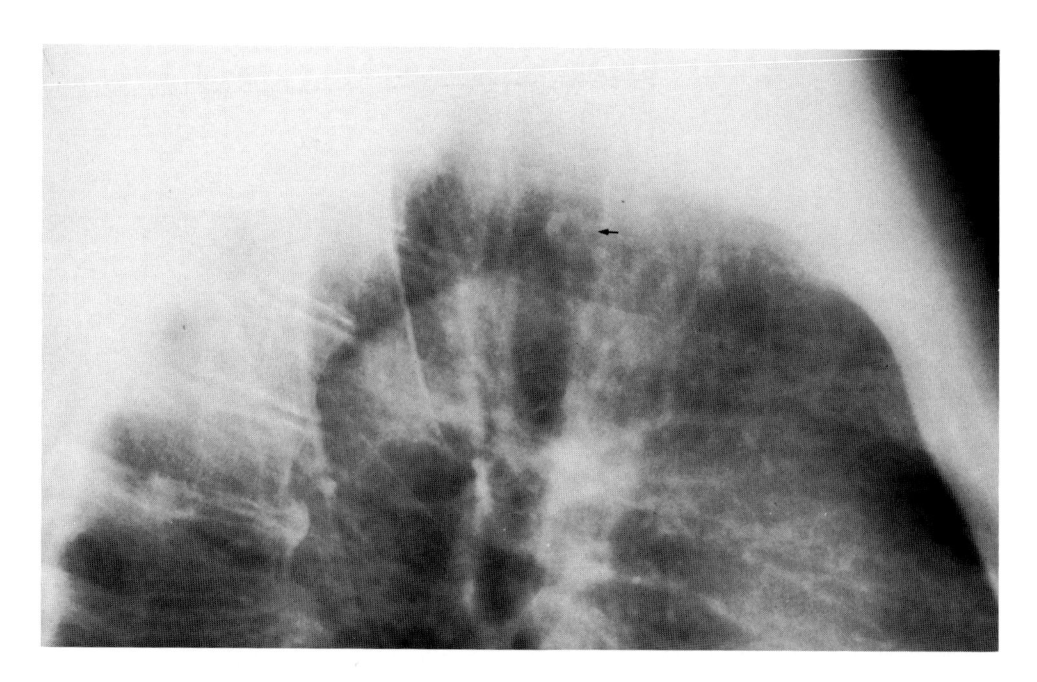

Fig. 4-3. An intratracheal soft tissue lesion, a papilloma (arrow).

1. Normal neonates and infants during expiration.
2. Tracheomalacia.
3. Localized areas of stenosis, whether congenital, postinflammatory, or postintubation (posttraumatic) (Fig. 4-4).
4. Pressure by an adjacent mass or vascular structure (Fig. 4-5). Vascular structures such as a double aortic arch, an aberrant subclavian artery, and a pulmonary sling (a left pulmonary artery arising from the right pulmonary artery) produce indentation on the posterior aspect of the trachea; an aberrant innominate artery or, occasionally, even a normal innominate artery and a cervical aortic arch produce an indentation on the anterior aspect of the trachea. Paratracheal mediastinal masses that can displace the trachea can also produce narrowing of the trachea.

Fig. 4-4. Tomographic section showing stenotic portion of the trachea (arrow).

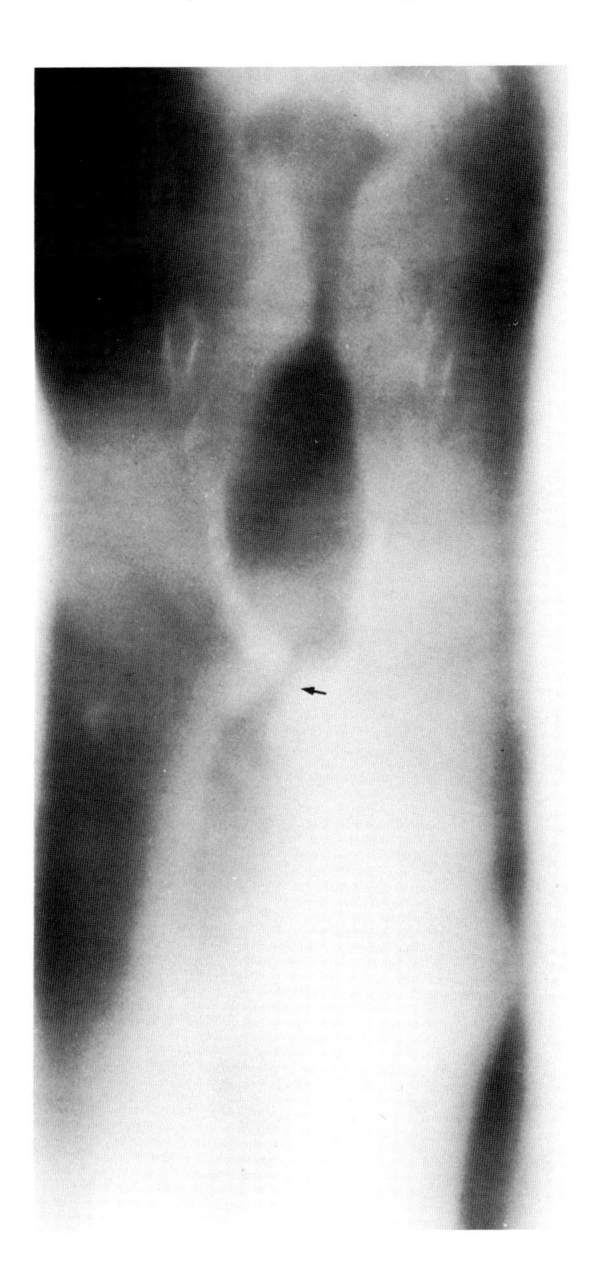

Fig. 4-5. *A.* Trachea (arrow) that is slightly displaced to the left by a substernal thyroid. *B.* Large substernal thyroid markedly displacing the trachea (arrow) to the left and narrowing it. *C.* Trachea (arrows) that has been displaced anteriorly by an aneurysm of the aorta.

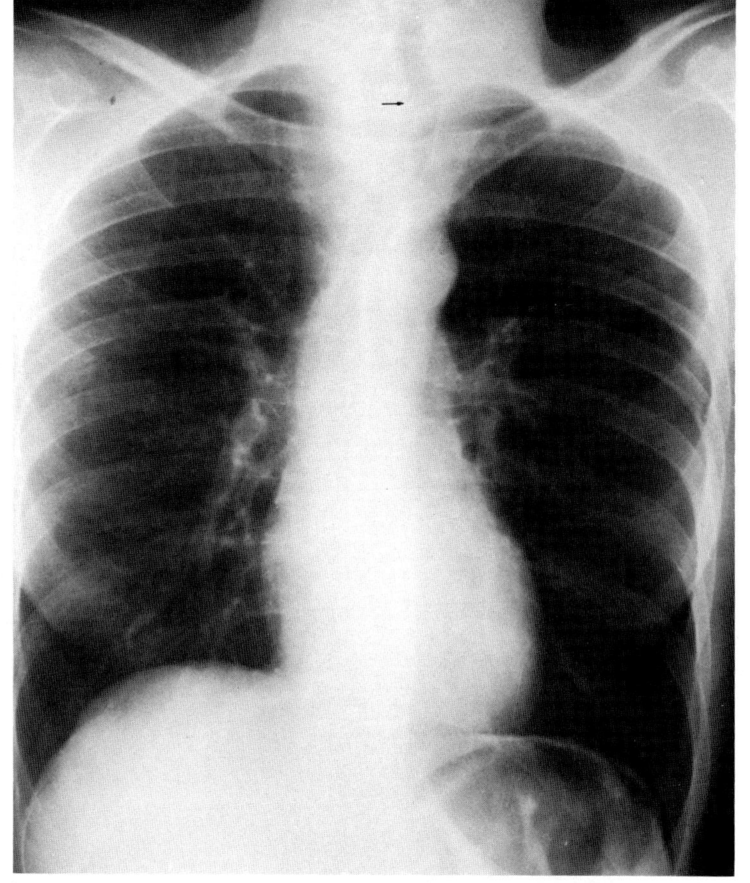

A

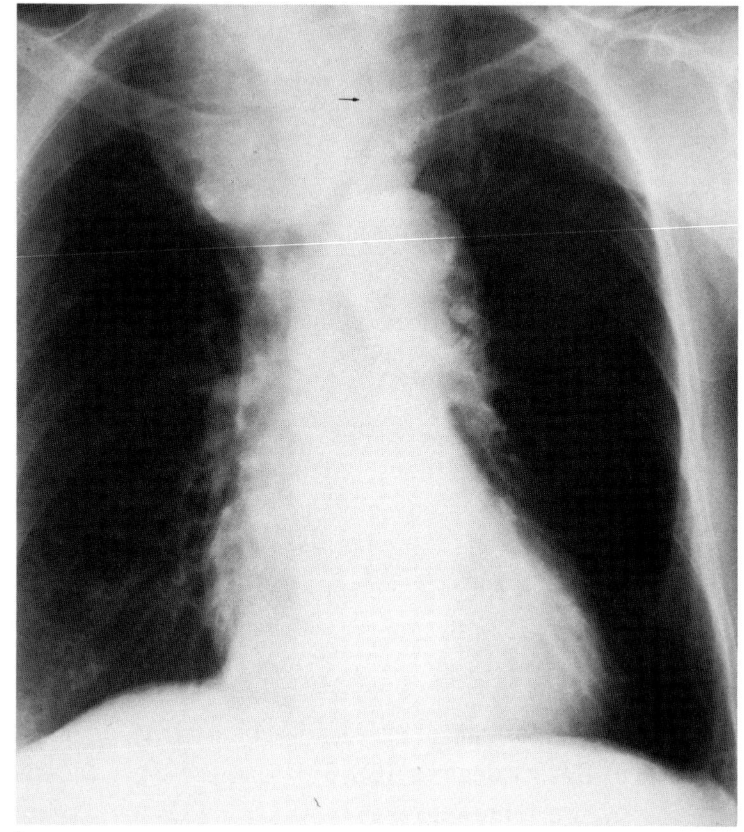

B

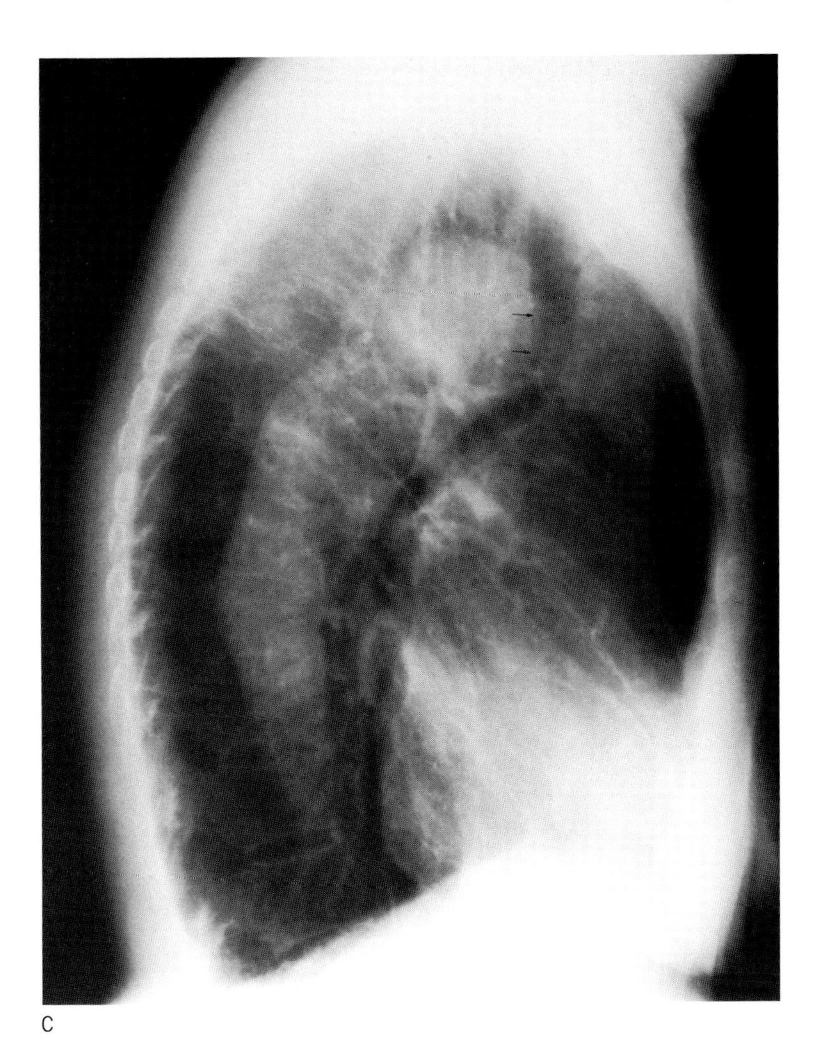

C

5. Inflammation resulting from
   a. Viral infection (croup or laryngotracheobronchitis) producing subglottic edema (steeple sign).
   b. Bacterial infection (extraglottic extension of *Haemophilus influenzae* supraglottitis, rhinoscleroma from *Klebsiella rhinoscleromatis,* and tuberculosis involvement from mycobacterial infection).
   c. Fungal infection (candidiasis, histoplasmosis, and mucomycosis all produce tracheal narrowing).
   d. Noninfectious inflammatory processes (Wegener's granulomatosis, sarcoidosis, relapsing polychondritis, amyloidosis, and tracheopathia osteoplastica).
6. Other causes such as tracheomalacial narrowing of the collapsed area and saber sheath trachea, the latter decreasing only the anteroposterior (AP) diameter of the trachea.

The carinal angle (the tracheal angle of bifurcation) is 35 to 87.50 degrees [60,61]. Carinal angle widening is produced by the following:

1. Enlarged left atrium and pulmonary veins.
2. Pericardial effusion.
3. Subcarinal mass or lymphadenopathy.

The carina is at the level of T3 in infants, T4-T5 in children, and T6 in adults.

In the lateral projection, two rounded lucencies are seen in the lower portion of the trachea. The upper lucency is formed by the right upper lobe bronchus; the lower, by the left upper lobe bronchus (Fig. 4-6). In the PA projection, only the right tracheal wall is well defined. It is normally 2 to 4 mm thick (Fig. 4-7A). In the lateral view, the posterior tracheal wall also normally measures 2 to 4 mm (Fig. 4-7B).

Thickening of the tracheal wall beyond 4 mm may be due to the following:

1. Abnormalities in the tracheal wall of mucosal, submucosal, or cartilaginus origin. These are the same entities that produce inflammatory narrowing of the trachea in both infectious and noninfectious diseases.

Fig. 4-6. *A.* Right upper lobe bronchus (arrow) and left upper lobe bronchus (arrowhead). *B.* Close-up of *A*. *C.* Right upper lobe bronchus (open arrow), left upper lobe bronchus (arrowhead), and bronchus intermedius (arrows). *D.* Close-up of *C*.

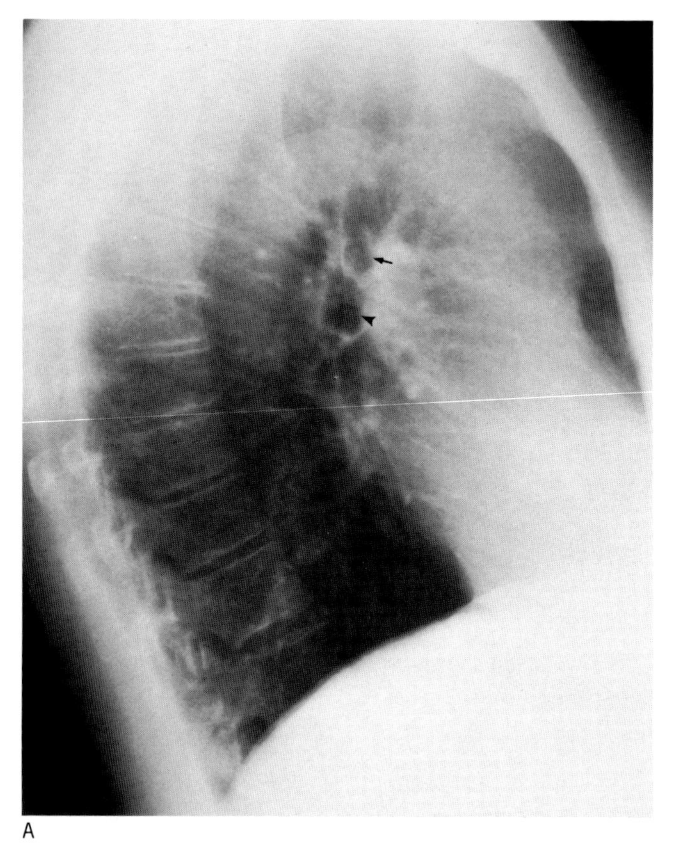

A

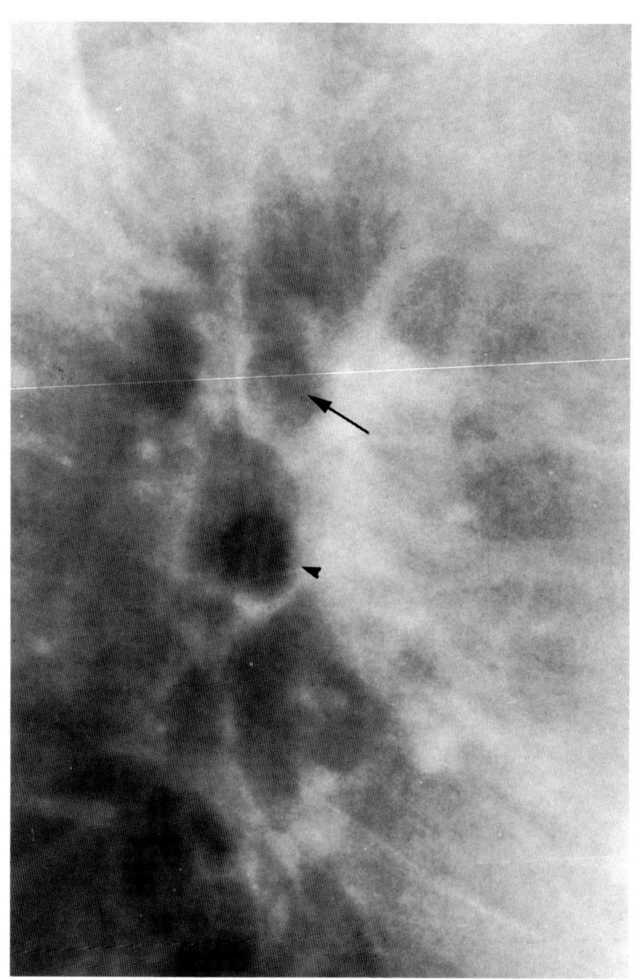

B

2. Mediastinal lymphadenopathy, specifically affecting the right paratracheal stripe or right tracheal wall thickness.
3. Pleural thickening, effusion, or hemorrhage, also affecting right tracheal wall thickening.
4. Esophageal carcinoma, which thickens the posterior tracheal wall (Fig. 4-8).

THE MEDIASTINUM

The mediastinum is the space between the two pleural sacs extending from the posterior aspect of the sternum to the front of the vertebral column, and from the thoracic inlet to the diaphragmatic surface. Anatomically, one can divide the mediastinum into the following sections (Fig. 4-9*A*):

1. Superior mediastinum. This portion may be separated from the lower two-thirds by a plane passing through the sternal angle in front to the fourth thoracic vertebra behind.
2. Anterior, middle, and posterior mediastinum. The lower

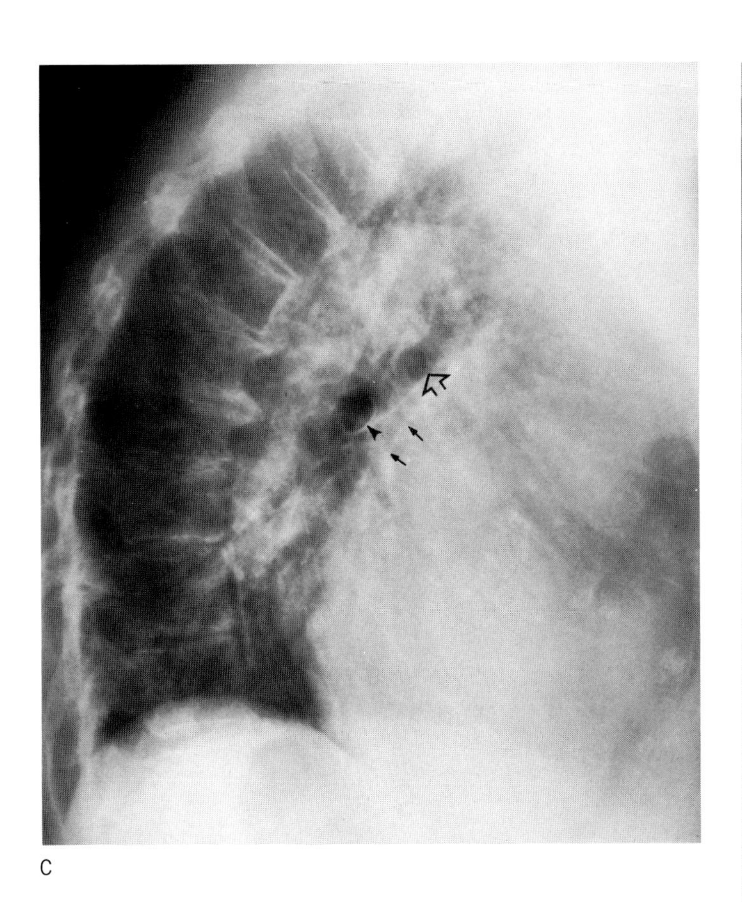

C

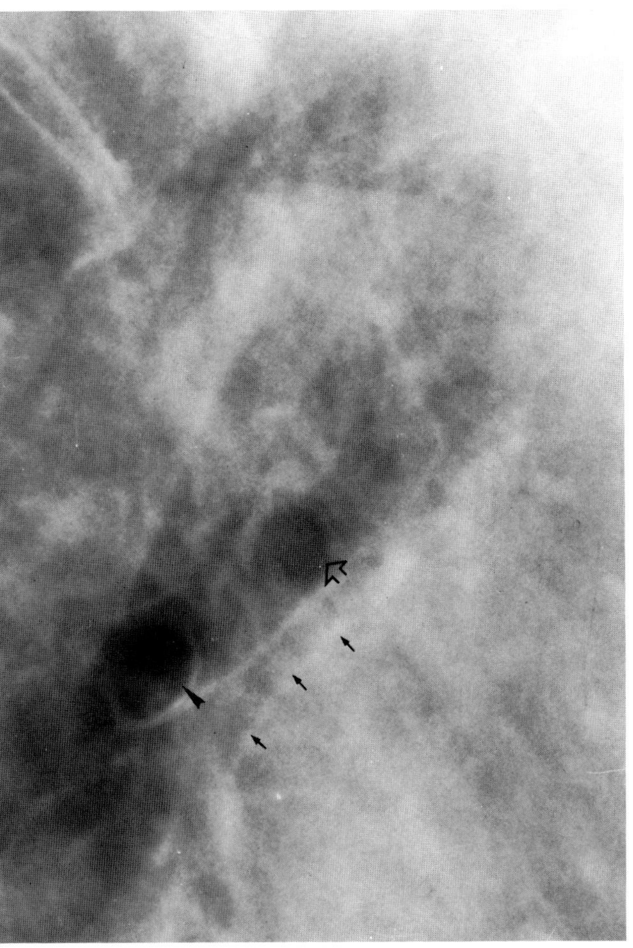

D

Fig. 4-7. *A*. Normal thickness of the lateral tracheal wall (arrows); normal azygous vein (arrowhead). *B*. Normal thickness of the posterior tracheal wall (arrows).

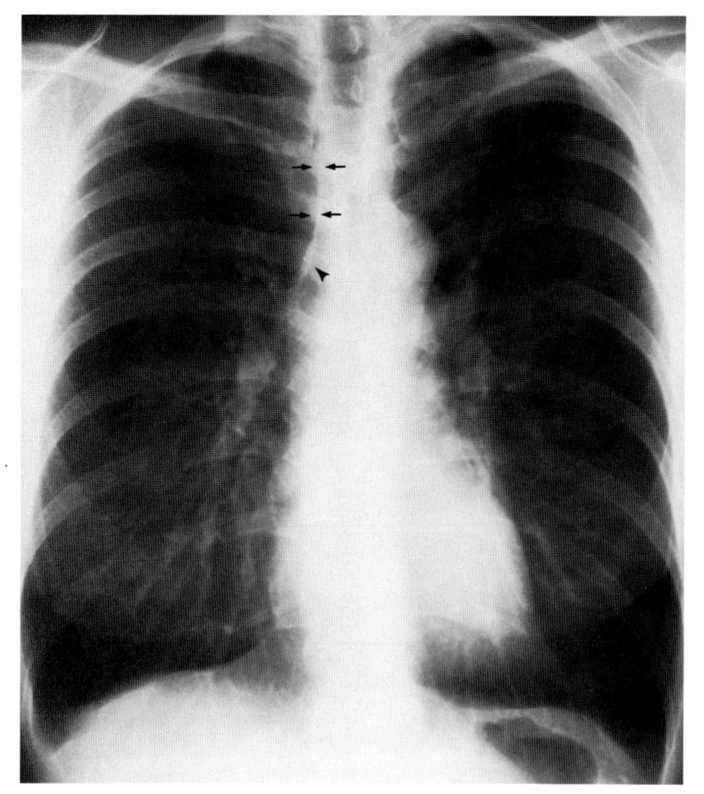

A

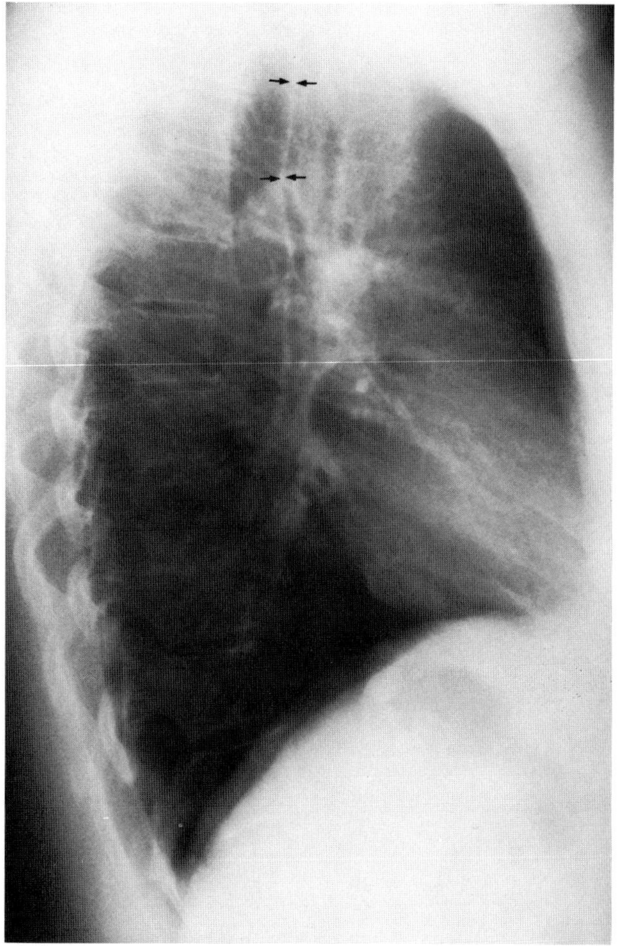

B

Fig. 4-8. Increased thickness of the posterior tracheal wall (arrows) in a patient with esophageal carcinoma. Contrast with the normal thickness in Fig. 4-7*B*.

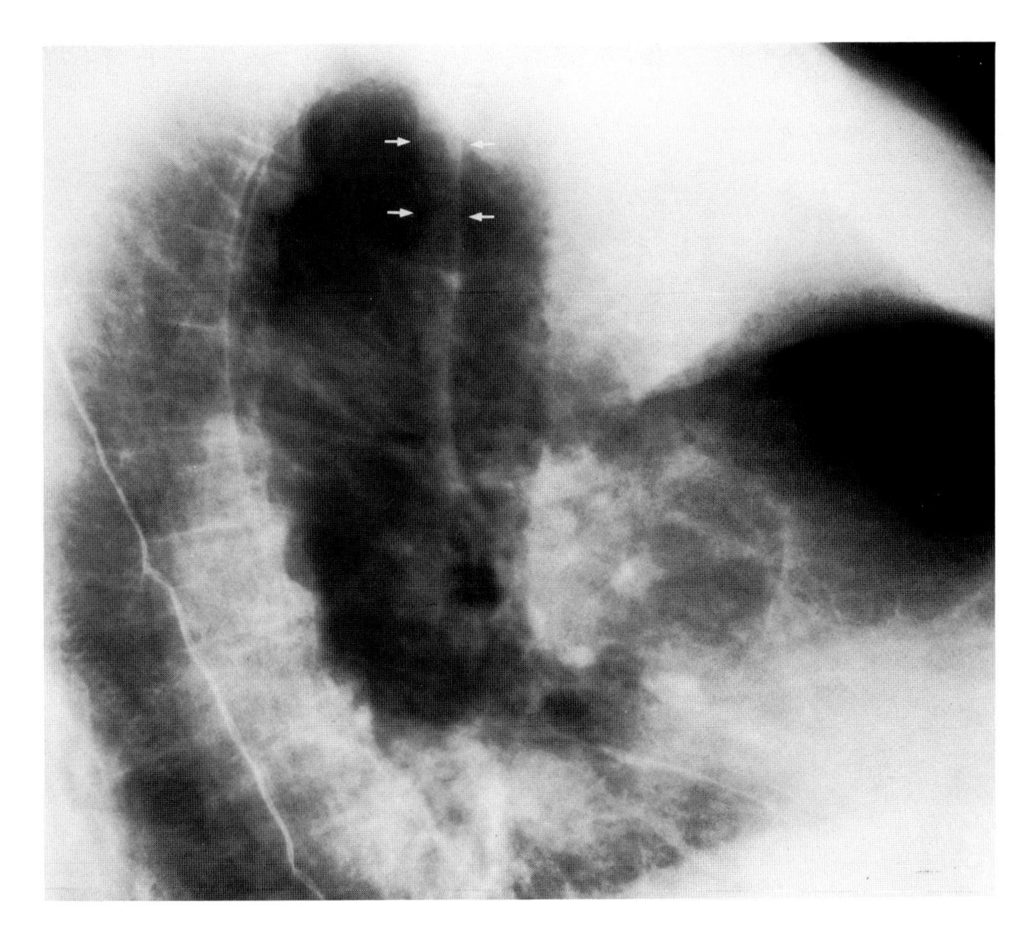

two-thirds of the mediastinum may then be divided into anterior, middle, and posterior compartments.

For radiologic purposes, it is not necessary to subdivide the mediastinum into superior and lower compartments, but it is helpful to divide it into anterior, middle, and posterior compartments. The most frequently used method of dividing the mediastinum into these three compartments is as follows (Fig. 4-9B):

1. Anterior. From the sternum anteriorly to the pericardium, aorta, and brachiocephalic vessels posteriorly.
2. Middle. From the pericardium to the posterior surface of the trachea.
3. Posterior. From the posterior surface of the trachea to the anterior surface of the vertebrae.

Fig. 4-9. *A.* Mediastinum divided into superior (4), anterior (1), middle (2), and posterior (3) compartments. *B.* Mediastinum divided into anterior (1), middle (2), and posterior (3) compartments. Anterior line placed slightly more posteriorly than where it should be and posterior line more anteriorly to allow the lines to be visualized by the contrast provided by the cardiac silhouette. *C.* Felson's division of the mediastinum into anterior (1), middle (2), and posterior (3) compartments.

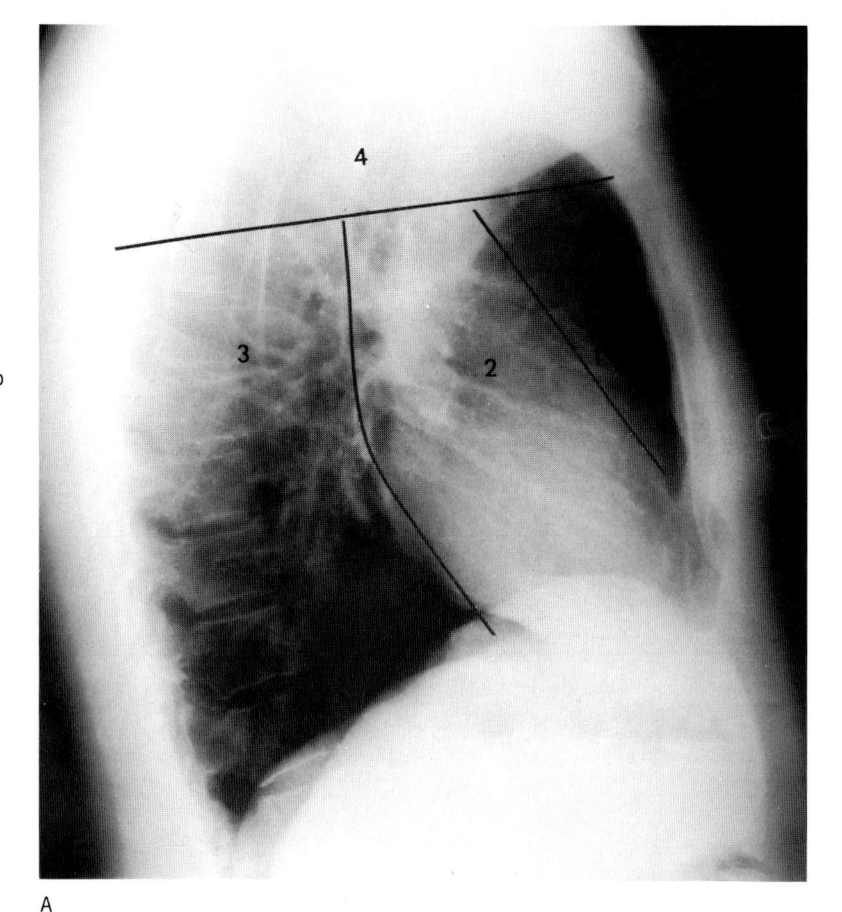

A

B

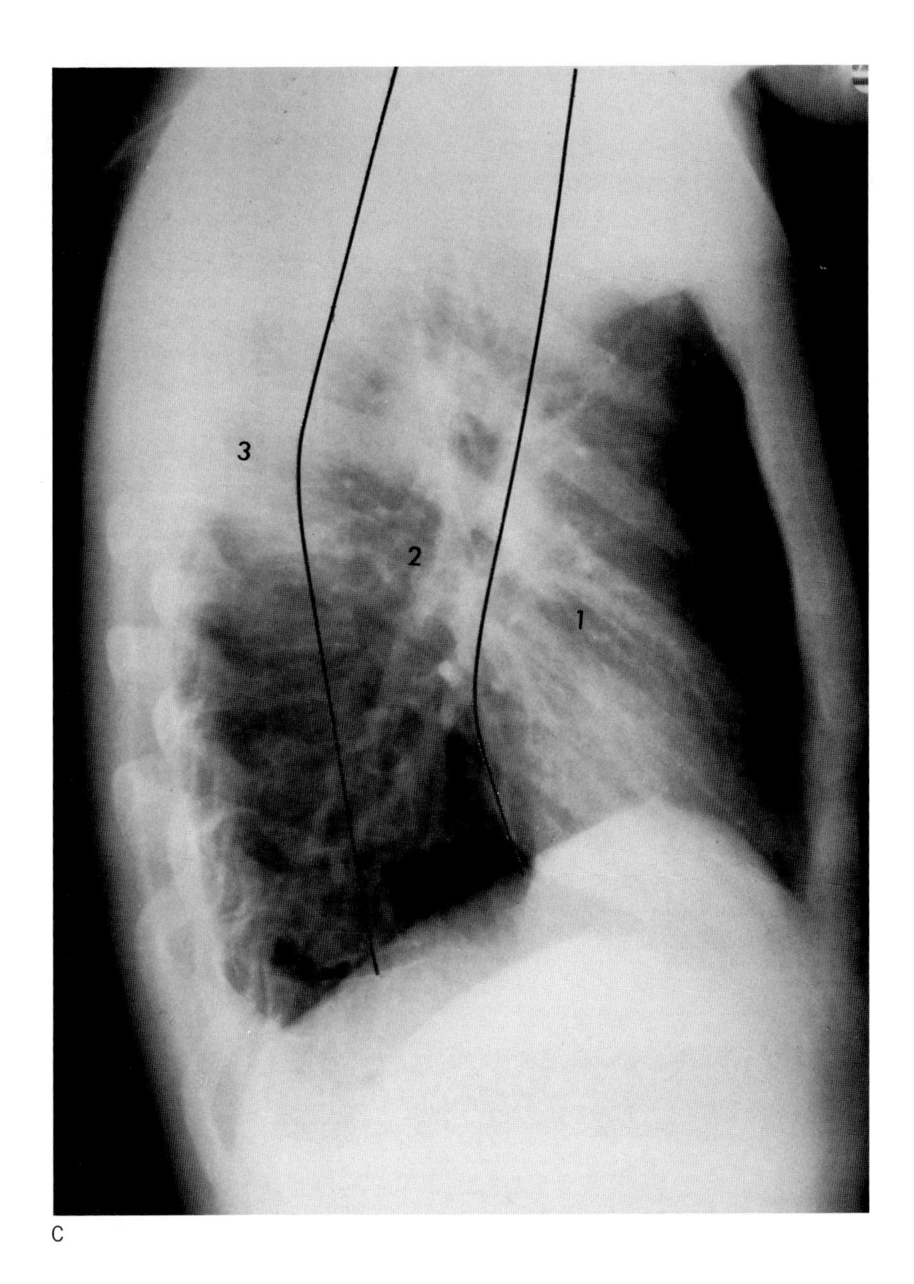

C

Felson [9], however, subdivides the mediastinum as follows (Fig. 4-9C):

1. Anterior. From the sternum to the anterior trachea, including the heart, into the anterior mediastinum.
2. Middle. From the anterior wall of the trachea to a plane 1 cm behind the anterior margin of the vertebral column.
3. Posterior. Posterior to the plane described in 2 above (Fig. 4-9C).

Because of the propensity of certain mediastinal masses to occur in one of the compartments, the consideration of the nature of a mass seen on an x-ray image will depend upon which method of subdivision one uses. Most surgeons prefer the second, more frequently used, method. The following is a list of lesions seen in each compartment using this method:

Anterior mediastinum—The "Ts" of the anterior mediastinum
- Thymoma
- Teratoma and germ cell tumors
- Thyroid and parathyroid lesions
- T-cell lesions (lymphoma) (Fig. 4-10)
- Tissues present—fibrous tissue (fibroma), blood vessel tumor (hemangioma), lymphatic vessels (lymphangioma), fat tissue tumor (lipoma)

Middle mediastinum
- Lymph node enlargement such as from lymphoma and sarcoidosis
- Morgagni hernia
- Pericardial lesions, cysts, fat necrosis, mesothelioma, tracheal or bronchial lesions such as bronchogenic cysts
- Cardiac lesions such as rhabdomyosarcoma
- Vascular lesions such as aneurysm of the aorta or dilatation of the main pulmonary artery, superior vena cava, and azygous vein

Posterior mediastinum
- Esophageal lesions such as duplication cyst (Fig. 4-11), diverticulum, carcinoma, and hiatus hernia
- Hernia through the foramen of Bochdalek
- Neurenteric and gastroenteric cysts
- Neurogenic tumors such as neurofibroma, neurilemmoma, schwannoma, ganglioneuroma, neuroblastoma, pheochromocytoma, chemodectoma, and meningocele

Pneumomediastinum is manifested radiologically as vertical streaks of lucency just lateral to the borders of the heart, with the parietal and visceral pleurae reflected by the lucent stripe (Fig. 4-12). Although this condition can be seen in the PA view, the lateral view (specifically, the cross-table lateral view) is more diagnostic.

Fig. 4-10. *A.* Lobulated anterior mediastinal mass (lymphoma) silhouettes out the cardiac border and simulates cardiomegaly. *B.* Anterior mediastinal mass (lymphoma) filling in the normal retrosternal clear space (arrows).

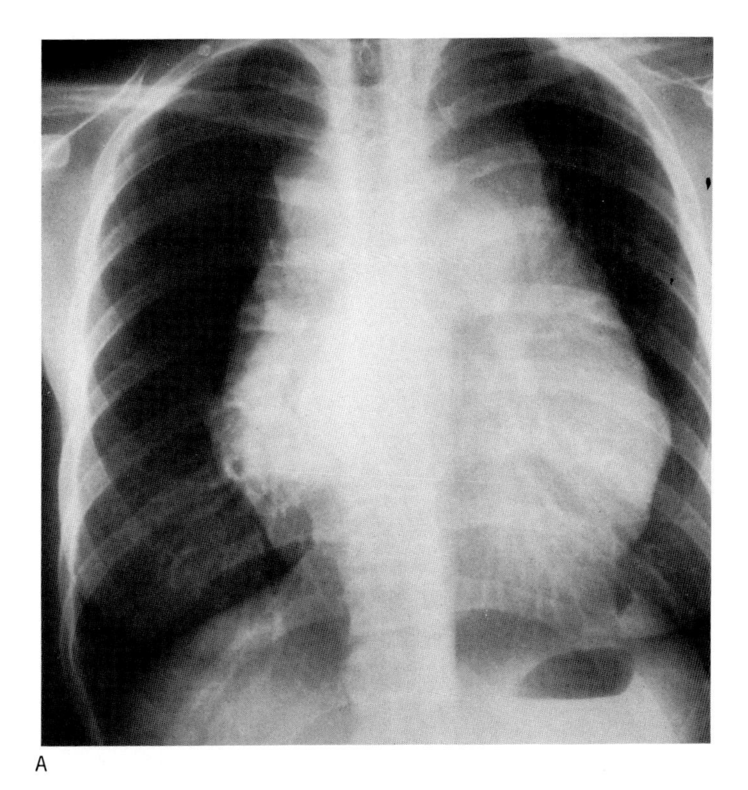

A

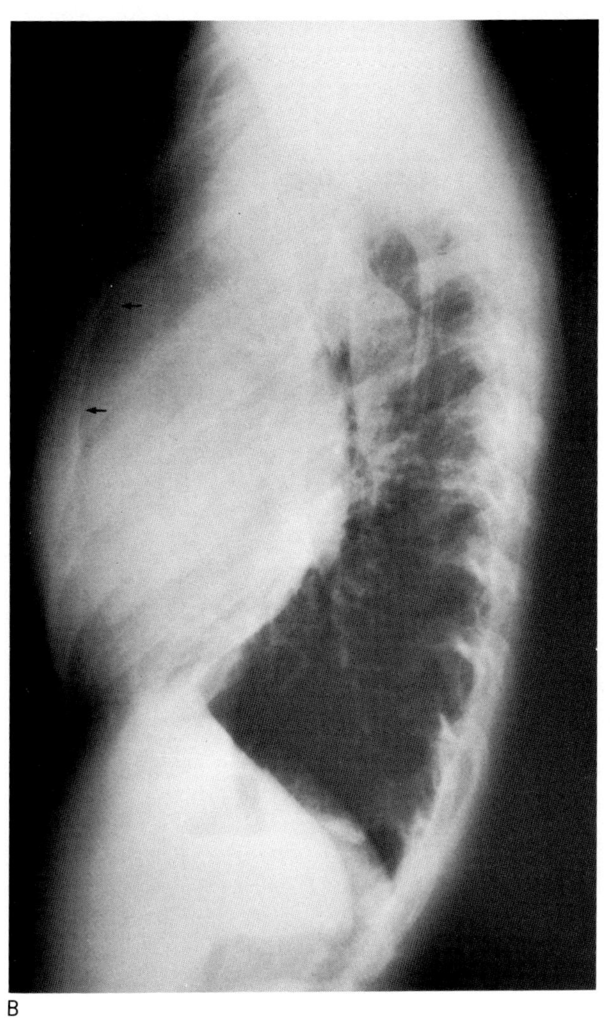

B

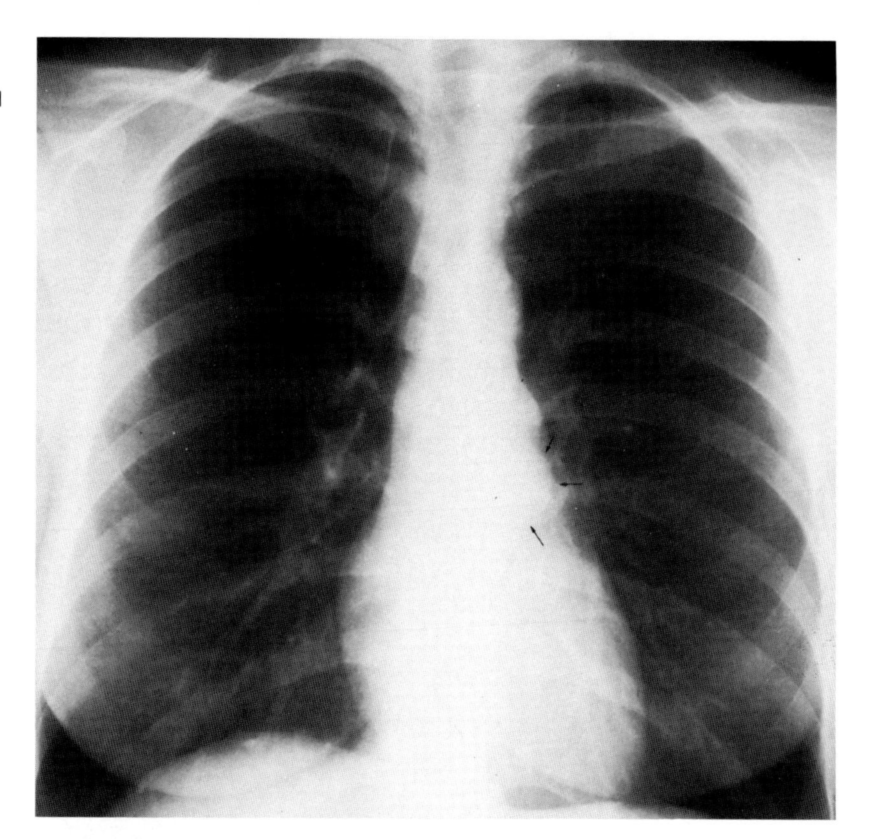

Fig. 4-11. Posterior mediastinal mass (arrows) seen through the cardiac silhouette, an esophageal duplicate cyst.

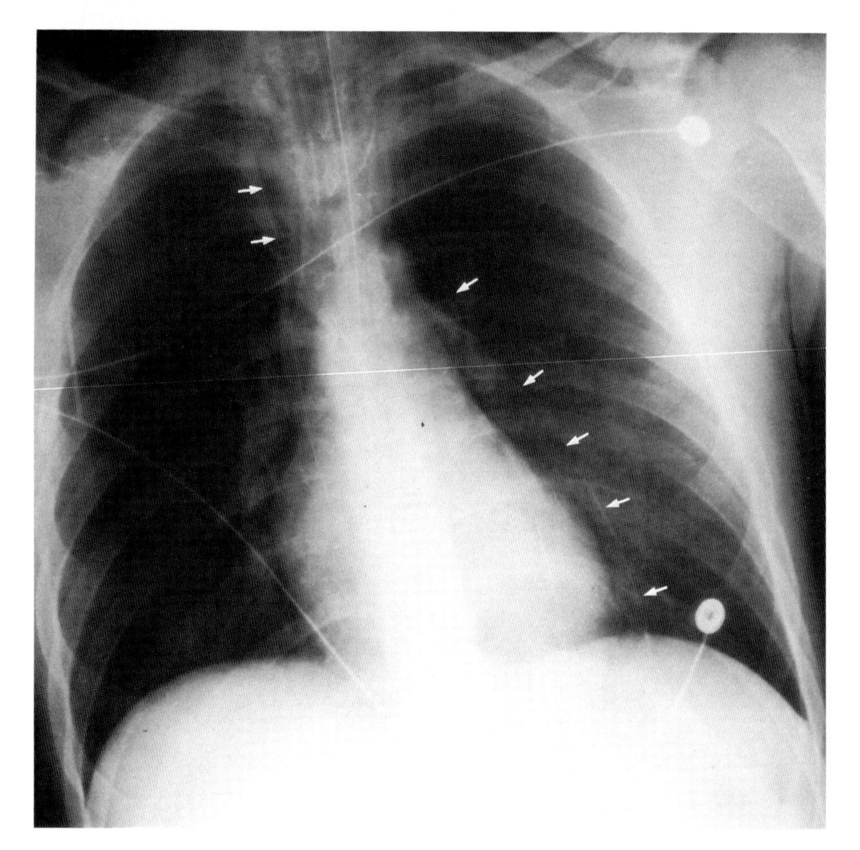

Fig. 4-12. Mediastinal pleura (arrows) that is laterally displaced by air within the mediastinum (pneumomediastinum).

Air can enter the mediastinum from a ruptured bronchus, trachea, or esophagus. It can also enter from the neck (especially during the course of tracheostomy or line placement when the negative pressure of the thorax draws air in through the incision), the retroperitoneum, and the lungs, in association with interstitial emphysema.

Small amounts of pneumomediastinum should be distinguished from the normal luceny of a *kinetic halo* around the heart. This artifactual halo is produced by normal cardiac motion. It is only moderately lucent and does not outline the pleural reflection. When air extends into the soft tissues of the neck or into the retroperitoneum, it is most likely secondary to a pneumomediastinum. Air in the mediastinum can readily enter the pleura, but pleural air does not enter an intact mediastinal pleura.

STEP 5 EVALUATE THE PLEURA

The normal pleura is usually not seen radiographically unless it is outlined by air on either side such as in the region of the fissures or in pneumothorax. Diseases of the pleura are manifested as (1) effusion, (2) thickening, (3) calcification, (4) masses, or (5) pneumothorax. Pleural masses have been previously discussed (see Pulmonary Nodules and Masses in Step 2). If the lines of pleural reflection are visible, they should be evaluated.

PLEURAL EFFUSION

The appearance of pleural effusion on a chest film depends upon the following factors:

1. Location of pleural fluid.
2. Amount of pleural fluid.
3. Projection of the chest radiograph.
4. Presence of air in the pleural space.
5. Presence of adhesions or fibrotic tissue.

Free pleural fluid in small amounts in the upright film may minimally blunt and opacify the costophrenic sulci, often with a visible meniscus (Fig. 5-1). The posterior costophrenic sulcus is the area first involved, followed by the lateral costophrenic sulci. Progressive increase in fluid leads to progressive opacification of the hemithorax with a rising meniscus level (Fig. 5-2) until there is total opacification of the involved side. Unless there is associated atelectasis on the same side, the mediastinum may shift to the contralateral side. The presence of air (pneumothorax) in the same side will produce an air-fluid level in place of the meniscus (Fig. 5-3).

Free fluid in the recumbent position produces a veil of haziness as it layers posteriorly on the involved side, progressing to total opacification with further increase in the amount of fluid. Apical capping is seen with moderate amounts of fluid in the supine position (Fig. 5-4).

Free fluid located in the subpulmonic region is characterized by a lateral peaking of the apparent diaphragmatic outline, in contrast to the normal central peak (Fig. 5-5).

Fig. 5-1. A meniscus (arrow) minimally blunting the R costophrenic angle in pleural effusion.

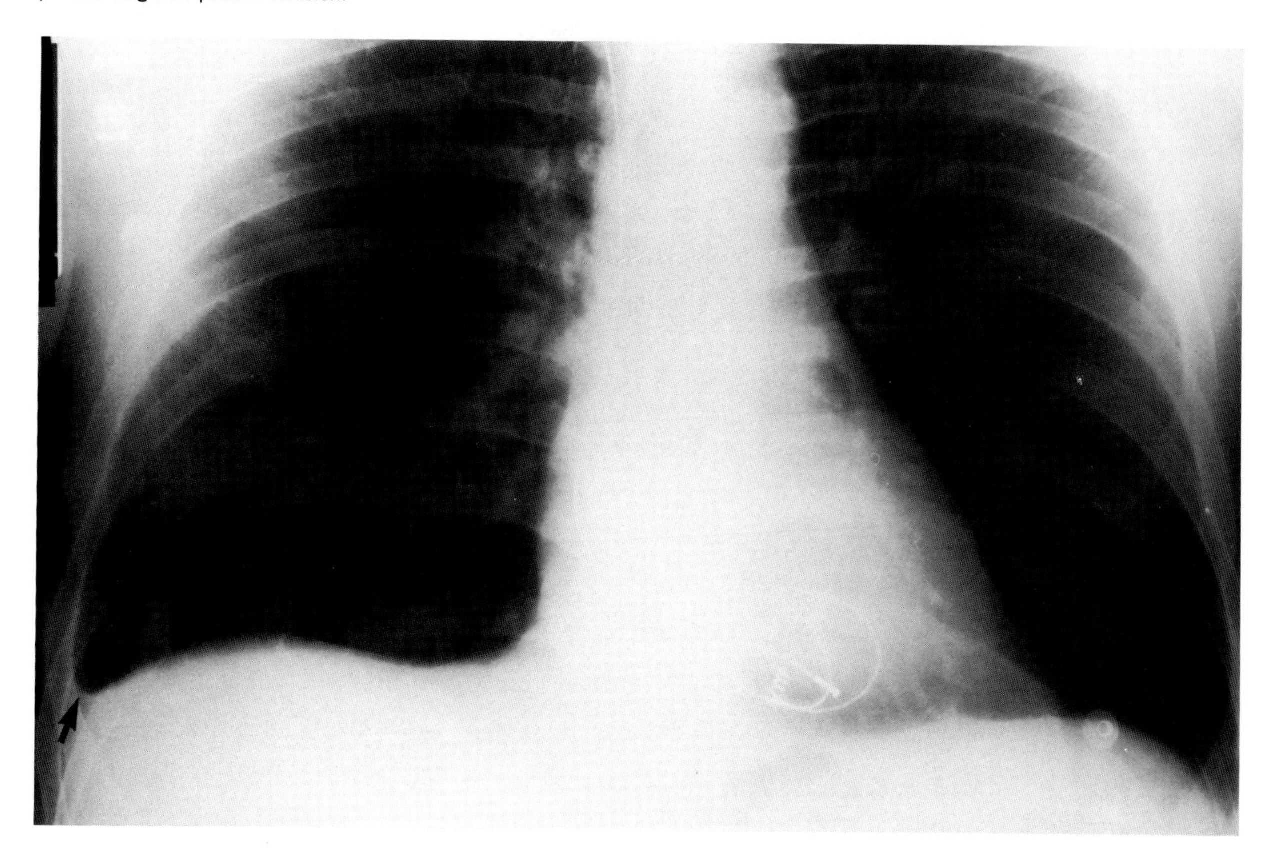

Fig. 5-2. *A*. Meniscus (arrows) formed by pleural effusion. *B*. Meniscus (arrowhead) produced by effusion in the posterior portion of the thorax.

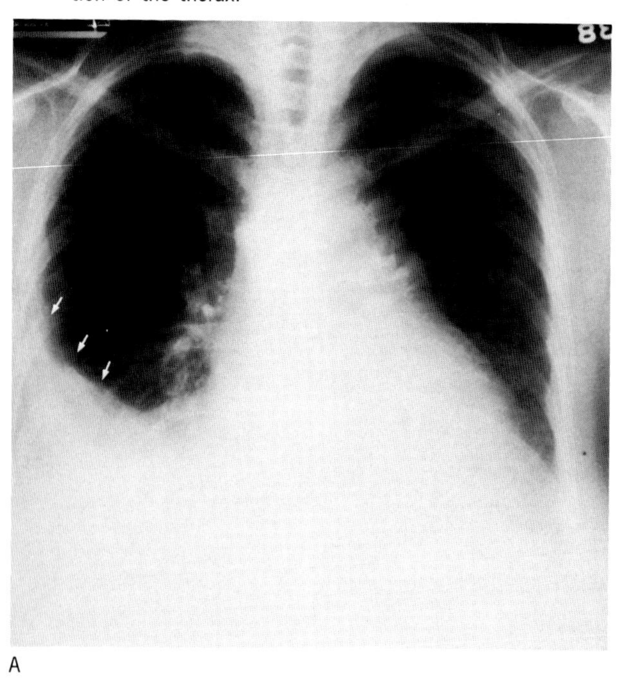

A

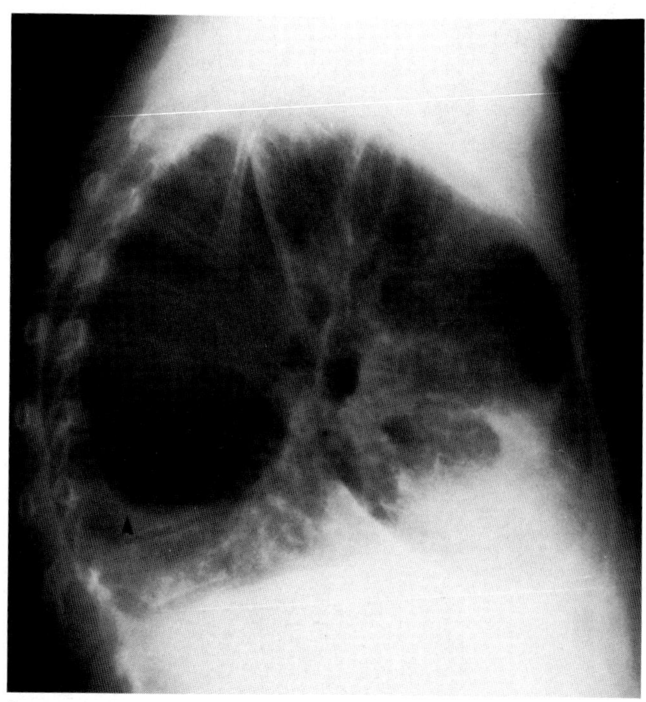

B

Fig. 5-3. Air-fluid level (arrow) in pneumohydrothorax.

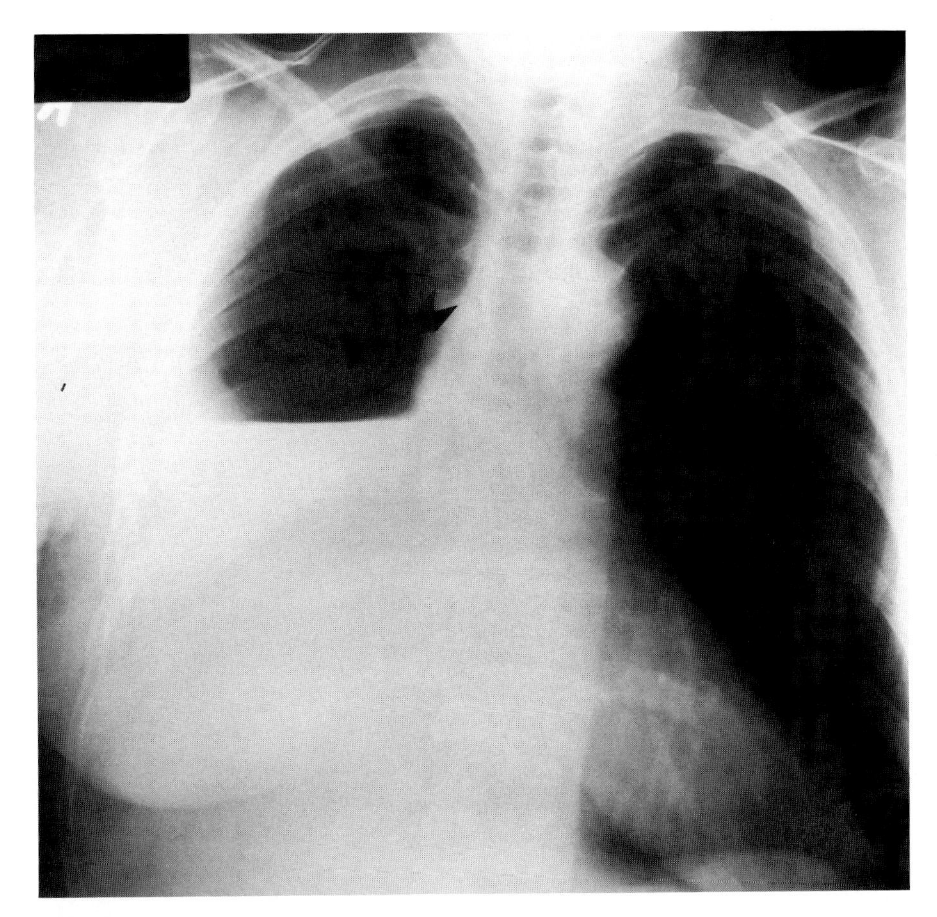

Fig. 5-4. Fluid in the apex in the supine position (arrowheads).

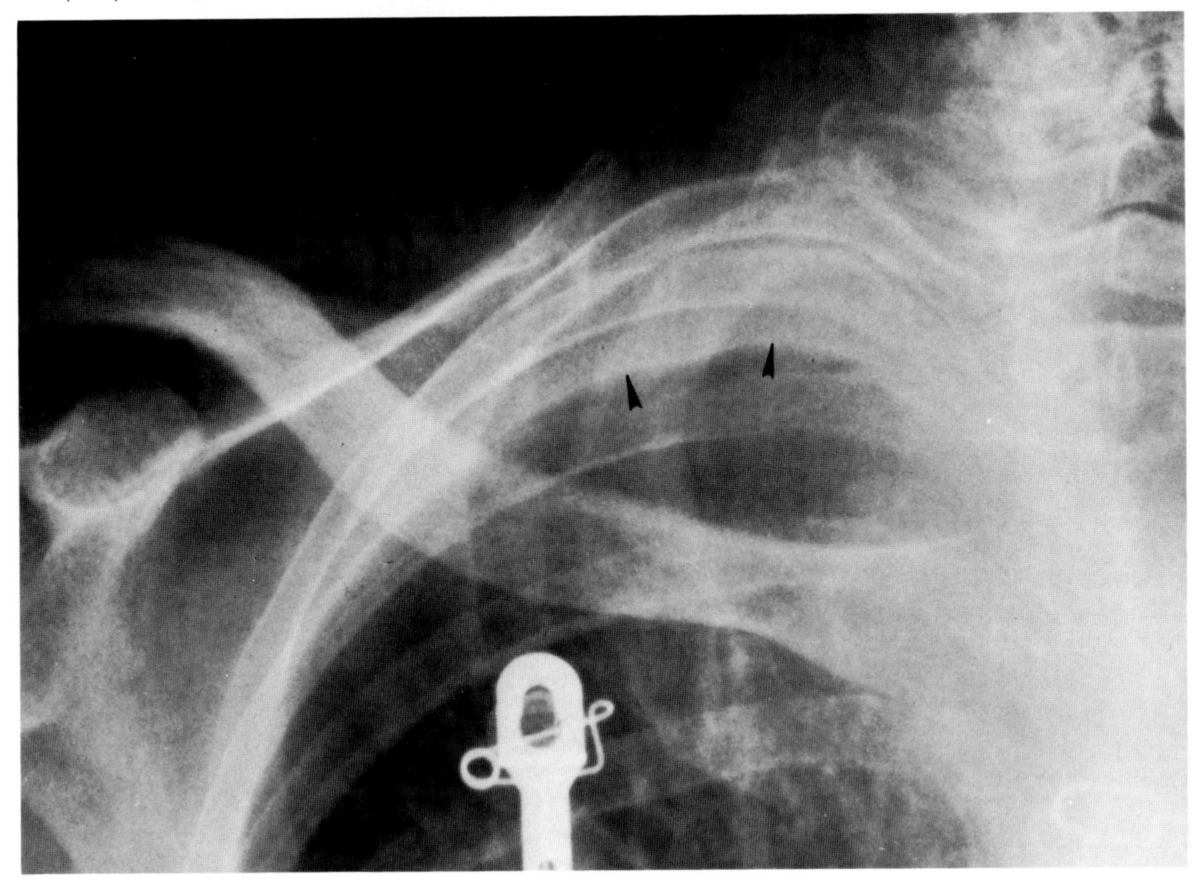

Fig. 5-5. *A.* PA film shows a more lateral peaking (arrows) of the right hemidiaphragm in subpulmonic effusion. *B.* Lateral decubitus view of the patient in *A* showing the moderately large volume of freely layering fluid (arrows). *C.* Elevated, slightly more lateral peak (arrow) of the apparent right hemidiaphragm. *D.* Lateral decubitus view shows large amount of R effusion (arrows) in patient in *C.*

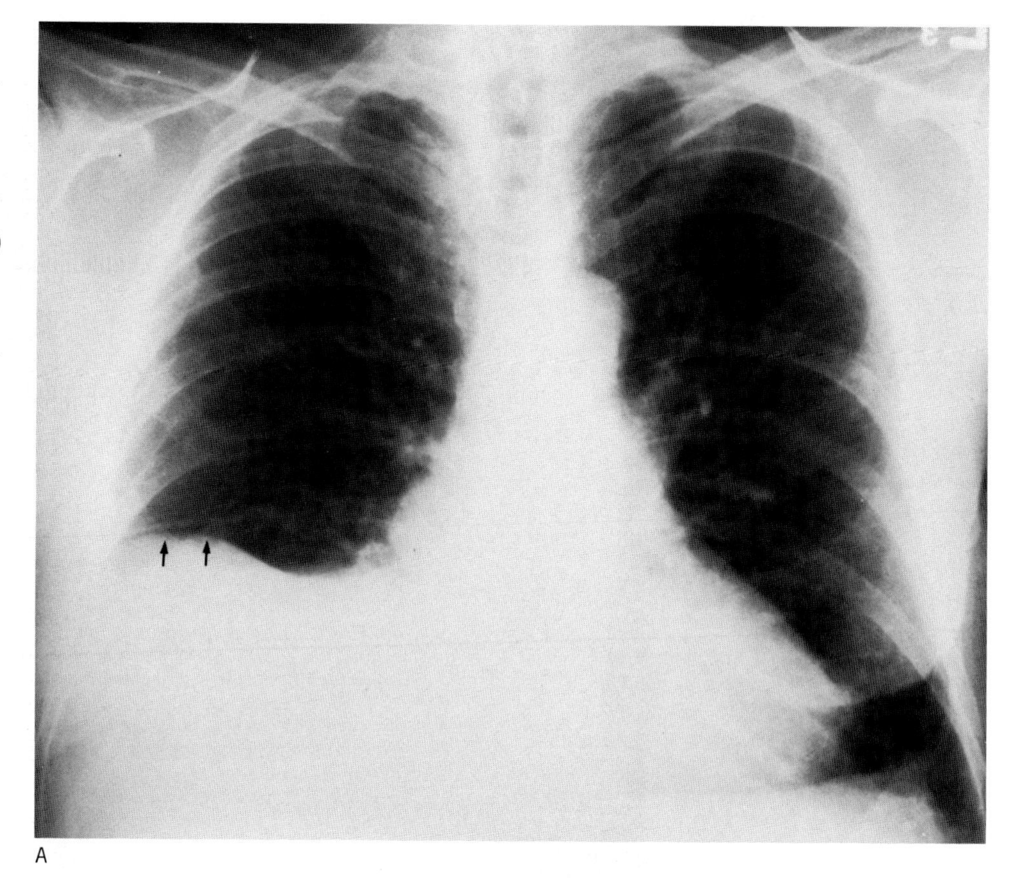

A

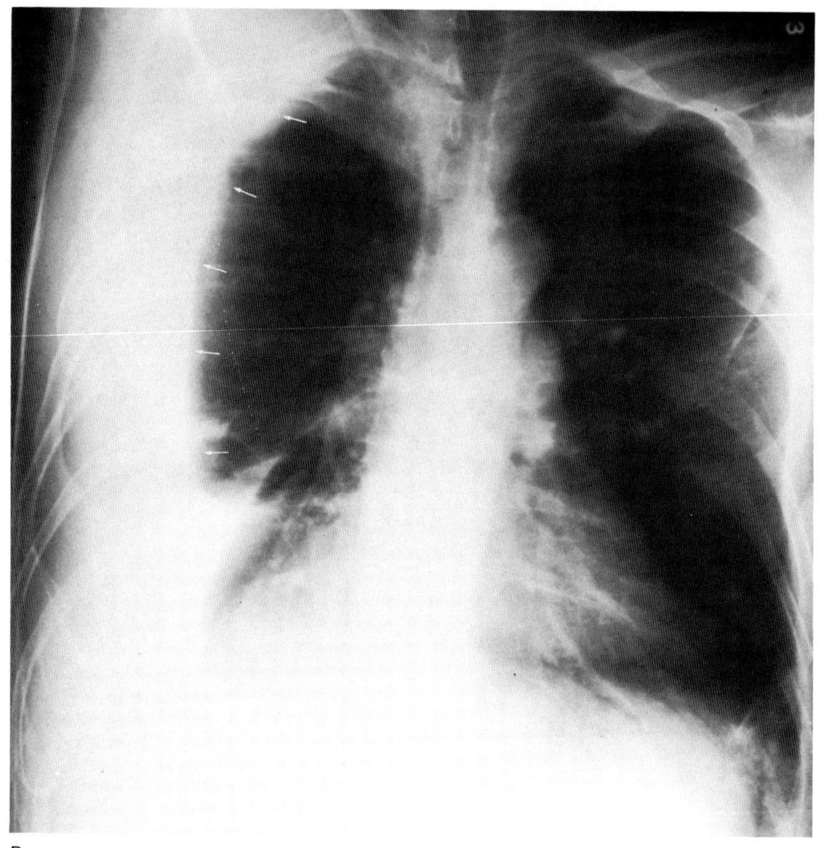

B

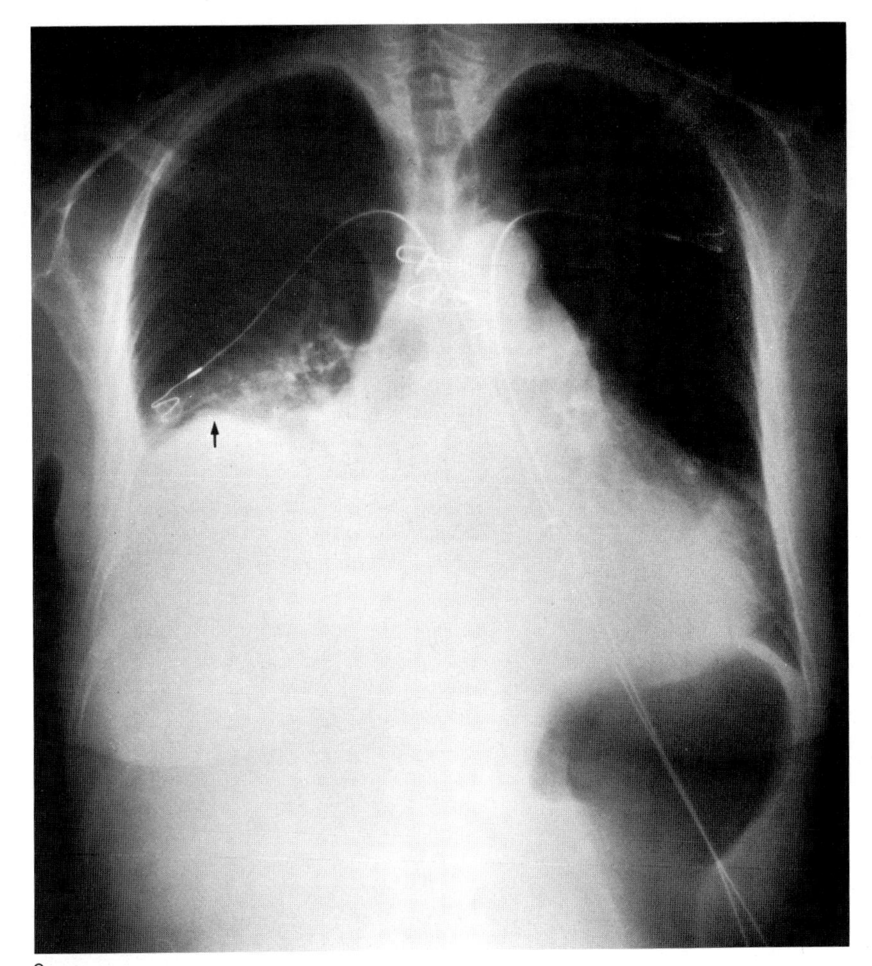
C

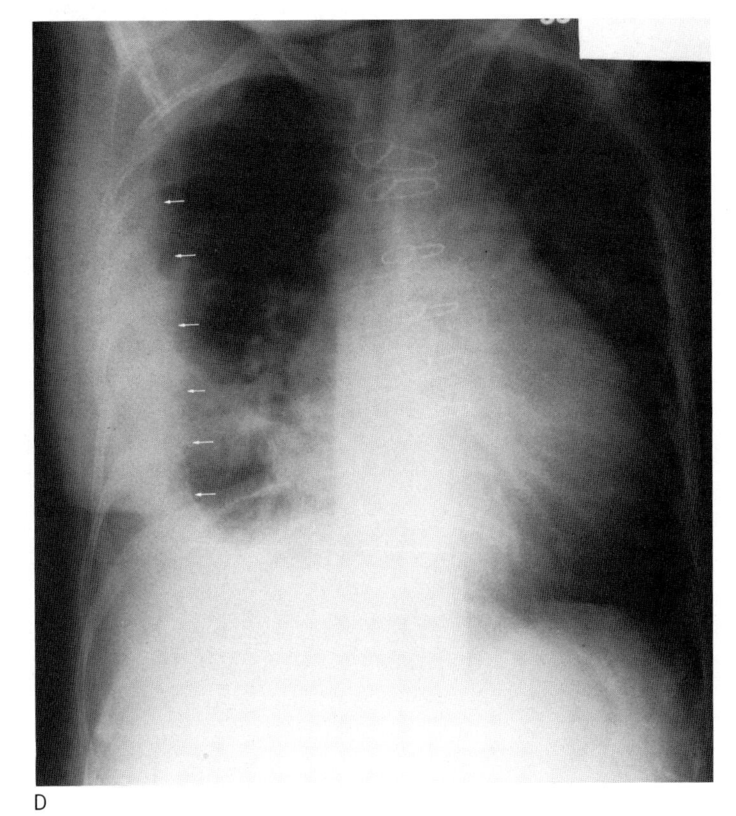
D

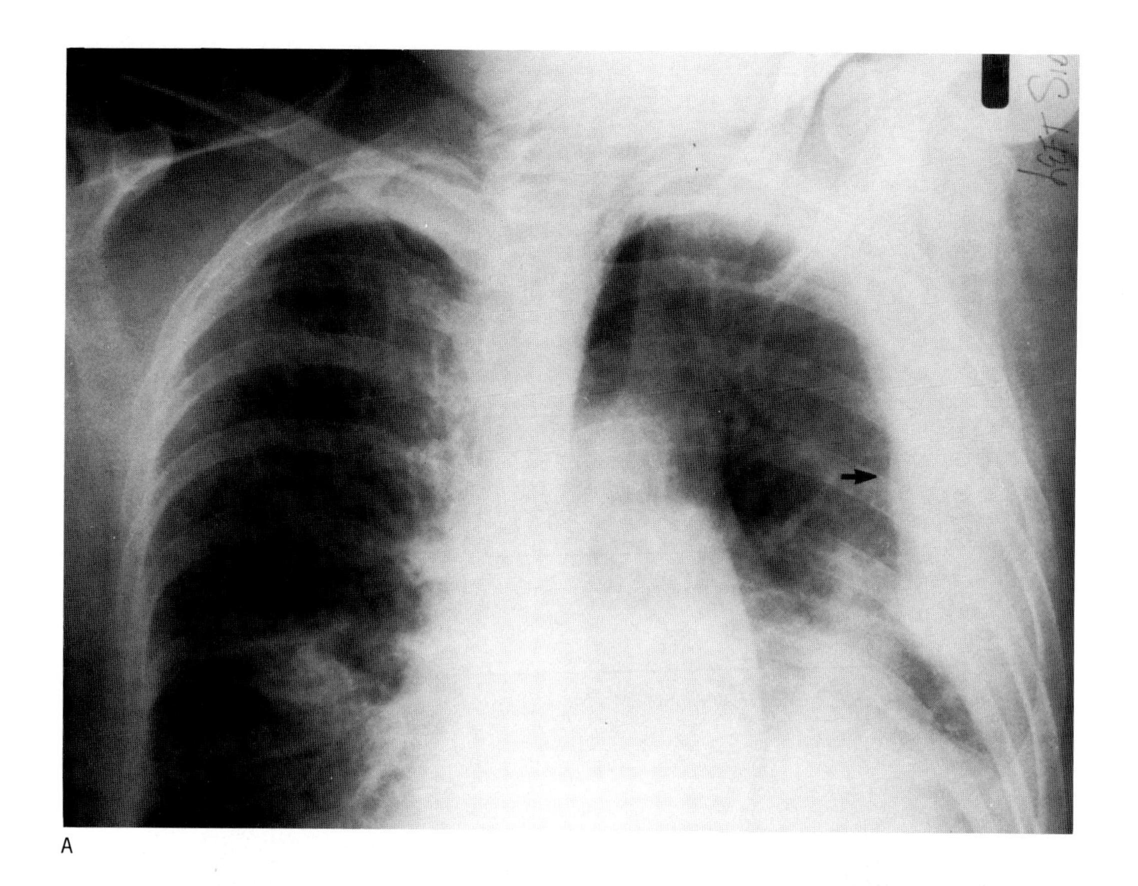

A

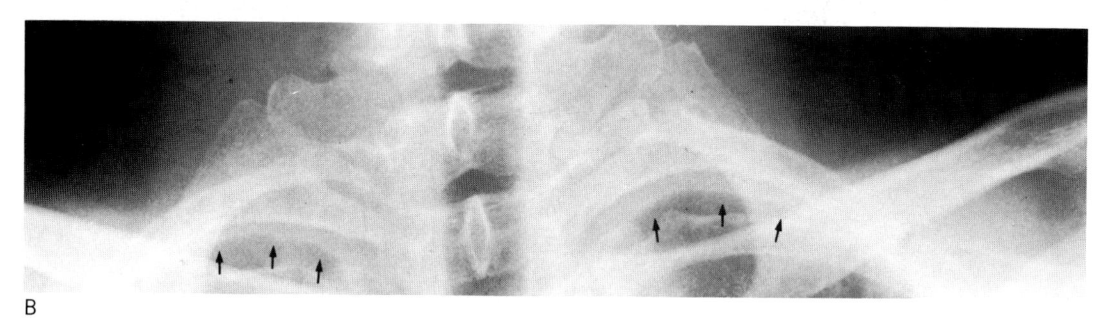

B

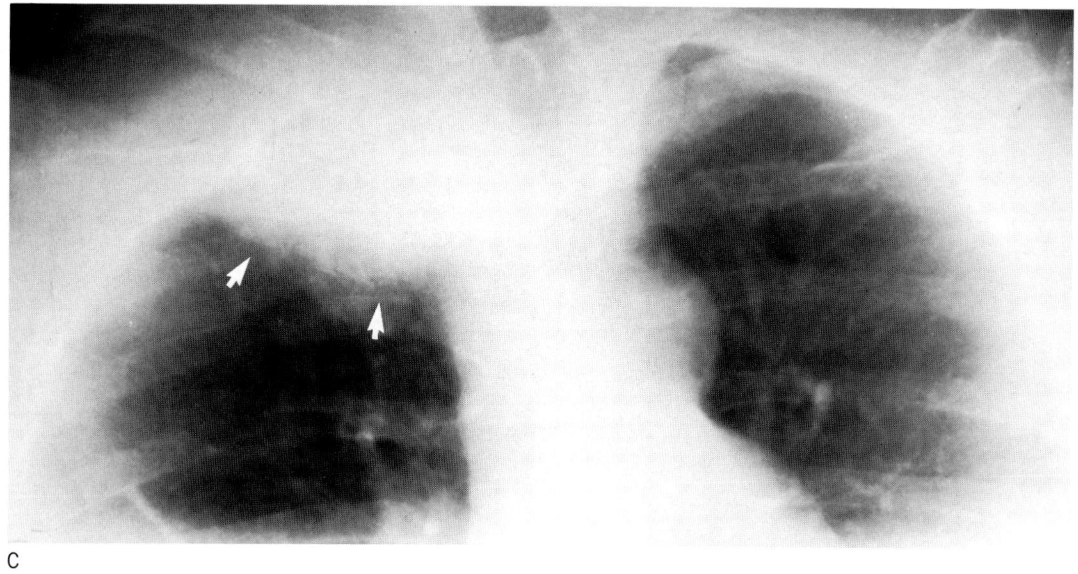

C

Fig. 5-6. *A*. Oblique view. Pleural thickening (arrow) in the left side. *B*. Smooth, symmetrical, bilateral, apical pleural thickening (arrows). *C*. Asymmetrical lobulated inferior border of a right apical density secondary to a Pancoast tumor (arrows).

Intrafissural pleural fluid produces a pseudotumor appearance with characteristic tapered edges (see Fig. 2-9).

In the decubitus view, a fluid level parallel to the dependent ribs forms (Fig. 5-5B and D), if not loculated. The decubitus view is helpful in differentiating between pleural thickening or loculated pleural fluid and free effusion.

Laboratory analysis of the fluid is necessary for diagnosis. Exudates have high protein (pleural fluid-serum ratio $\geqslant$ 0.5), high lactate dehydrogenase (LDH) ($>$ 200), high LDH ratio (pleural fluid-serum ratio $\geqslant$ 0.6), and high cellular content. Bloody fluid suggests neoplasm, infarction, or trauma. A high white-cell count suggests an infection. In tuberculosis, the predominant cells are lymphocytes, and the glucose value is low. Rheumatoid effusion also has very low glucose levels that do not rise with a challenge of intravenous glucose, because of impaired glucose transport. Rheumatoid factor and low C3 and C4 complement are also seen in rheumatoid effusion. An elevated amylase suggests pancreatitis, or more rarely, a neoplasm, infection, or esophageal rupture, when the source of the amylase is the salivary gland.

A pleural density may actually be an empyema. Empyema may be distinguished from a lung abscess by assessing the width of the fluid level on two views. Pleural location is suggested when there is a discrepancy in the diameter of the air-fluid level between the two 90-degree projections, such as in the PA and lateral views. Diameter of the air-fluid level in an intrapulmonary lesion (i.e., abscess) is the same in both views.

A change in the contour of the air-fluid level with a change in position (e.g., from supine to decubitus) also suggests pleural location [62].

PLEURAL THICKENING, ADHESION, AND CALCIFICATION

Pleural thickening is recognized as a rim of increased density parallel to the chest wall, between lung and ribs (Fig. 5-6). When seen *en face,* it varies from a faint haze over the lung to a thick density with rather sharp, angular margins. Oblique views best show pleural thickening. To distinguish from pleural effusion, especially when the costophrenic sulcus is involved and blunted, a decubitus view is necessary. If the density at the costophrenic sulcus is convex toward the lung, loculated fluid or the Hampton's hump of pulmonary infarction should be ruled out. Apical pleural thickening is often postinflammatory. It is seen in 1 to 1.2 percent of healthy individuals, is bilateral in one-third of the cases, and is more frequently seen on the left [9]. If the apical pleural thickening is unilateral, thick, and lumpy, Pancoast's tumor should be excluded clinically.

Pleural adhesion is recognizable by its effect on adjacent lung or mediastinal structures and is accompanied by pleural thickening. Adhesions at the diaphragmatic surfaces should be distinguished from a juxtaphrenic peak [63] that occurs in upper lobe or, occasionally, in lower lobe collapse, and from the inferior pulmonary ligament. The juxtaphrenic peak is seen in the medial two-thirds of the diaphragm, anteriorly. The inferior pulmonary ligament is seen posteriorly behind the cardiac silhouette. Pleural calcification is seen as a very dense line parallel to the chest wall or, if *en face,* as a thick density over the lung (Fig. 5-7). Often it appears lacelike *en face* (Fig. 5-8), because of the irregularity of calcium deposition.

Fig. 5-7. *A* and *B* are PA and lateral views of multiple calcified pleural plaques simulating pulmonary nodules.

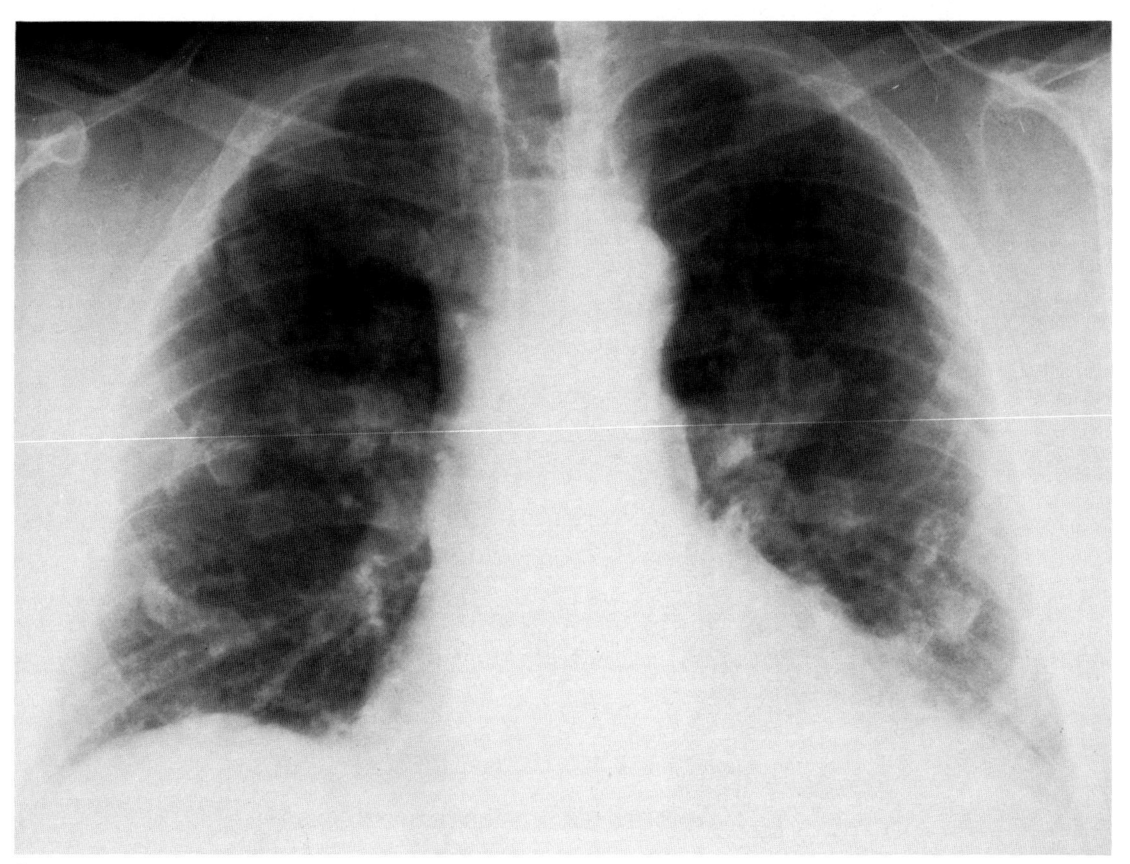

A

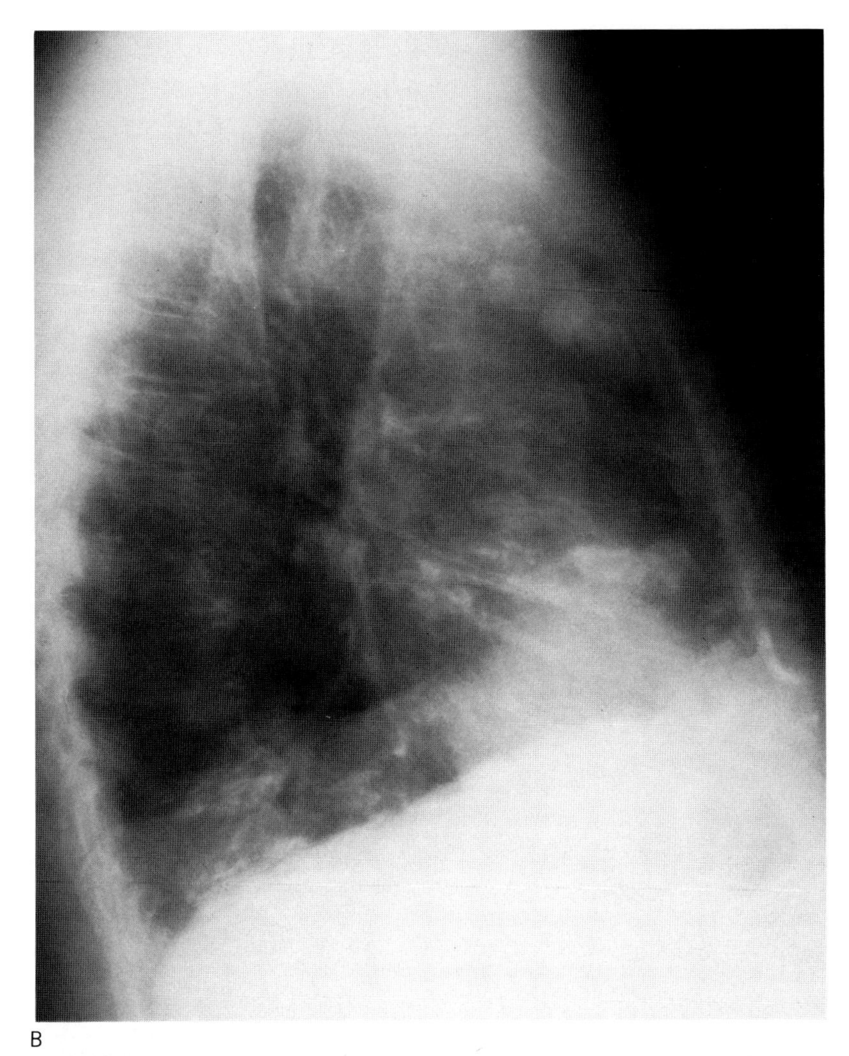

B

Fig. 5-8. Pleural calcification producing a lacelike pattern (white arrow) in the right lower lobe in a patient with asbestos-related pleural abnormality. Note faint diaphragmatic calcifications bilaterally (black arrows).

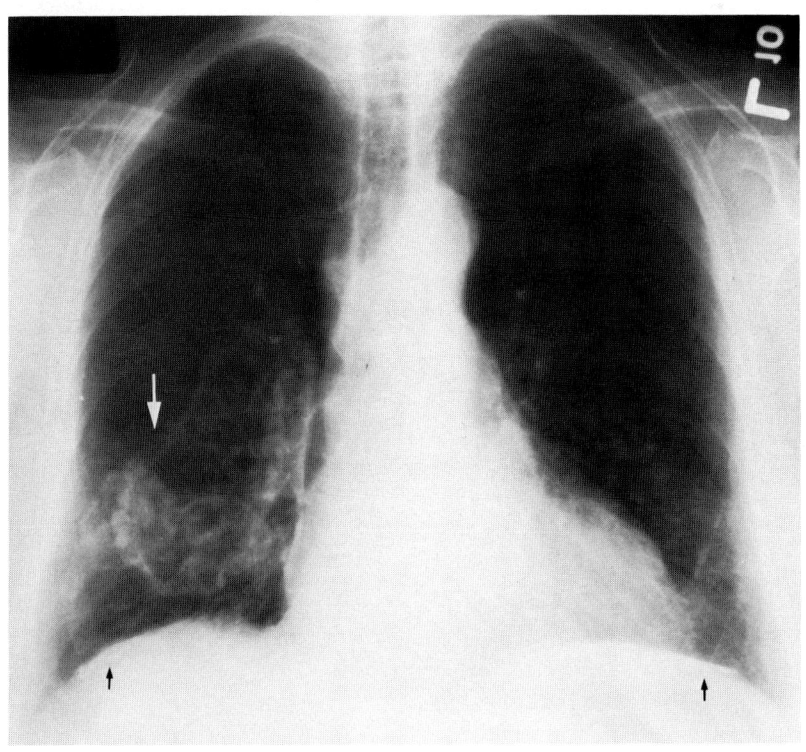

PNEUMOTHORAX

Pneumothorax is recognized as a zone of radiolucency devoid of lung markings between the lung and the thoracic wall. The pleura outlined by air on both sides becomes visible. The lung partially (Fig. 5-9A and B) or wholly (Fig. 5-9C) collapses and drops to the most dependent position, suspended by its fixed attachment at the pulmonary ligament. The density of the partially collapsed lung may not increase when compared with the opposite side because blood flow through it diminishes correspondingly, the degree of diminution of flow progressing with increasing collapse. Thus, the ratio of air to blood is maintained, and the lung density remains unaltered [64].

As air accumulates in the pleura, the mediastinum tends to shift to the opposite side. This is best seen in a film taken during the expiratory phase of respiration. For the mediastinum to shift, it is not necessary for the intrapleural pressure to become positive on the side of the pneumothorax, but merely less negative. If the mediastinum is not fixed, the diminished negative pressure on the side of the pneumothorax creates enough of an imbalance between the pleural pressures

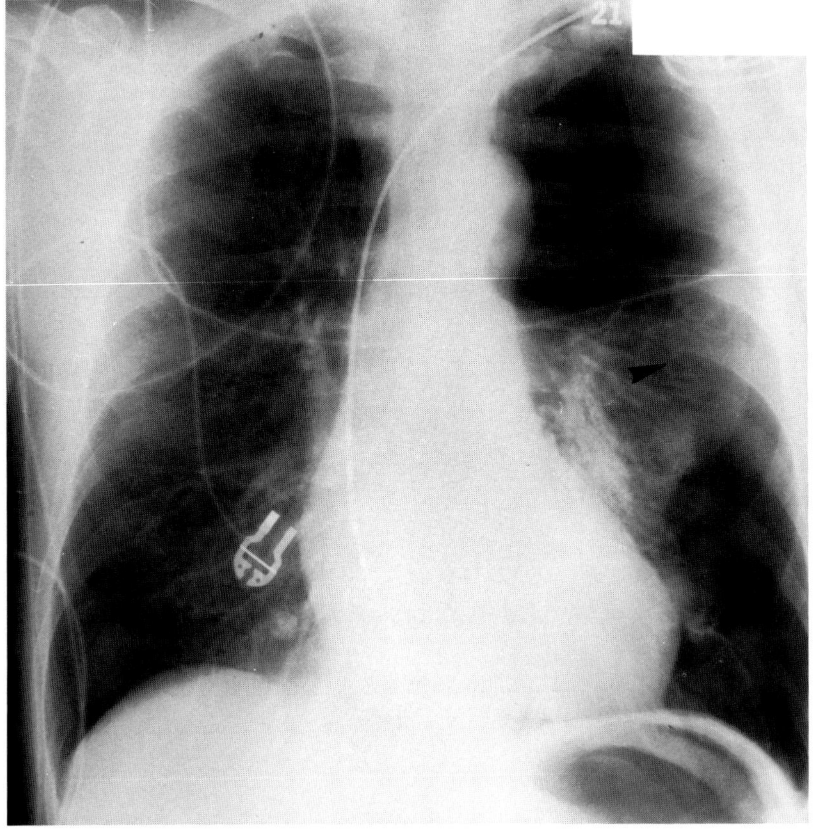

A

Fig. 5-9. *A.* Partially collapsed left lung (arrowhead). Note density of lateral portions of collapsed lung is similar to noncollapsed lung since air-to-blood ratio is maintained. *B.* Pneumothorax on the right side. Note that the stiff lung (arrowhead) from ARDS does not collapse as in *A* or *C*. *C.* Collapsed lung (arrowhead) in pneumothorax. Note mediastinal shift to the right.

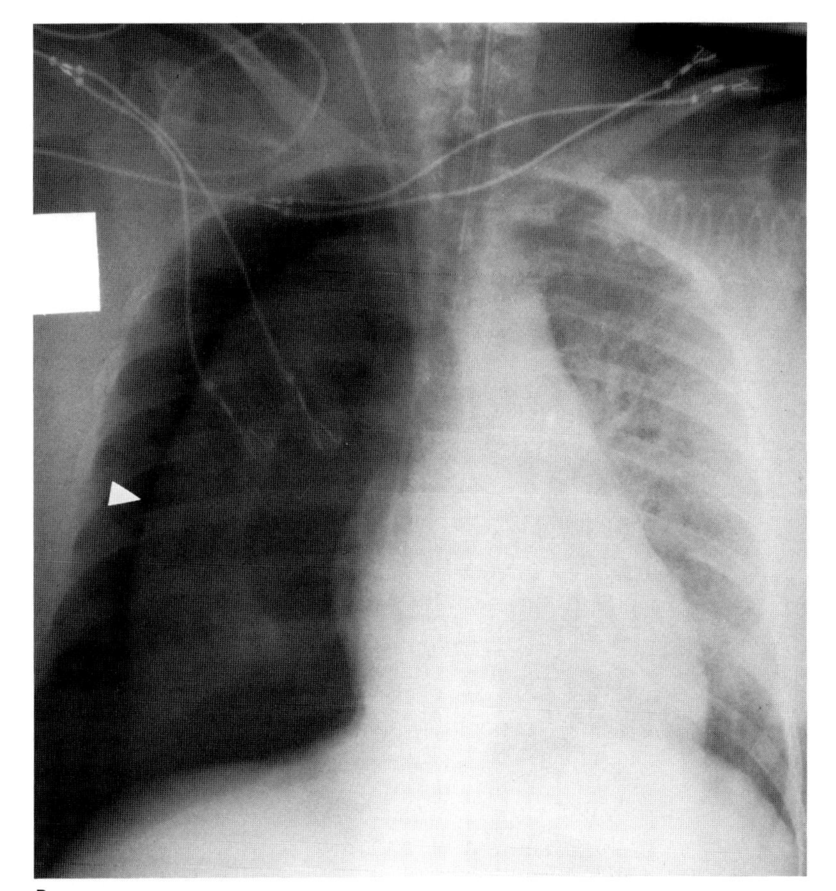
B

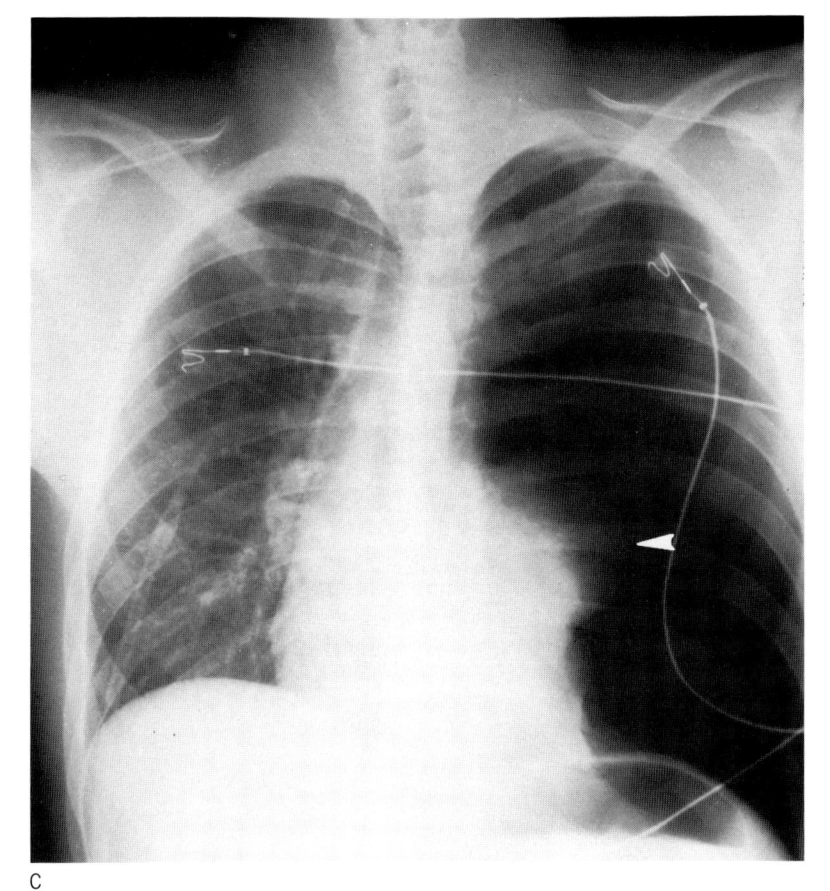
C

of the two sides to cause mediastinal displacement during the expiratory phase of respiration.

If the mediastinum is not fixed, tension pneumothorax will cause a shift of the mediastinum to the opposite side during both inspiratory and expiratory phases of respiration. In addition, flattening, with progression to reversal of the normal curve of the hemidiaphragm, occurs in tension pneumothorax. Rhea and associates [65] described a simple, reproducible means of measuring the percentage of pneumothorax present in upright PA and lateral films. This measurement is accomplished by taking an average interpleural distance and calculating the percentage of pneumothorax from it using the total lung volume of the partially collapsed lung and the total hemithoracic volume as parameters. For prediction of pneumothorax size based on average intrapleural distance, a nomogram (Fig. 5-10) is available to estimate easily the size of the pneumothorax.

The distribution of air in the pleural cavity is affected by pleural adhesions and by disease of the underlying lung. Adhesions prevent lung retraction, and extensive adhesions may therefore lead to a loculated pneumothorax. A diseased lung, especially one with scarring or atelectasis secondary to bronchial obstruction, tends to retract to a greater degree than the adjacent lung. Obstructive emphysema, consolidation, and interstitial emphysema make the lung rigid and interfere with retraction, keeping the lung or the involved seg-

Fig. 5-10. Rhea's nomogram for calculating the percentage of pneumothorax.

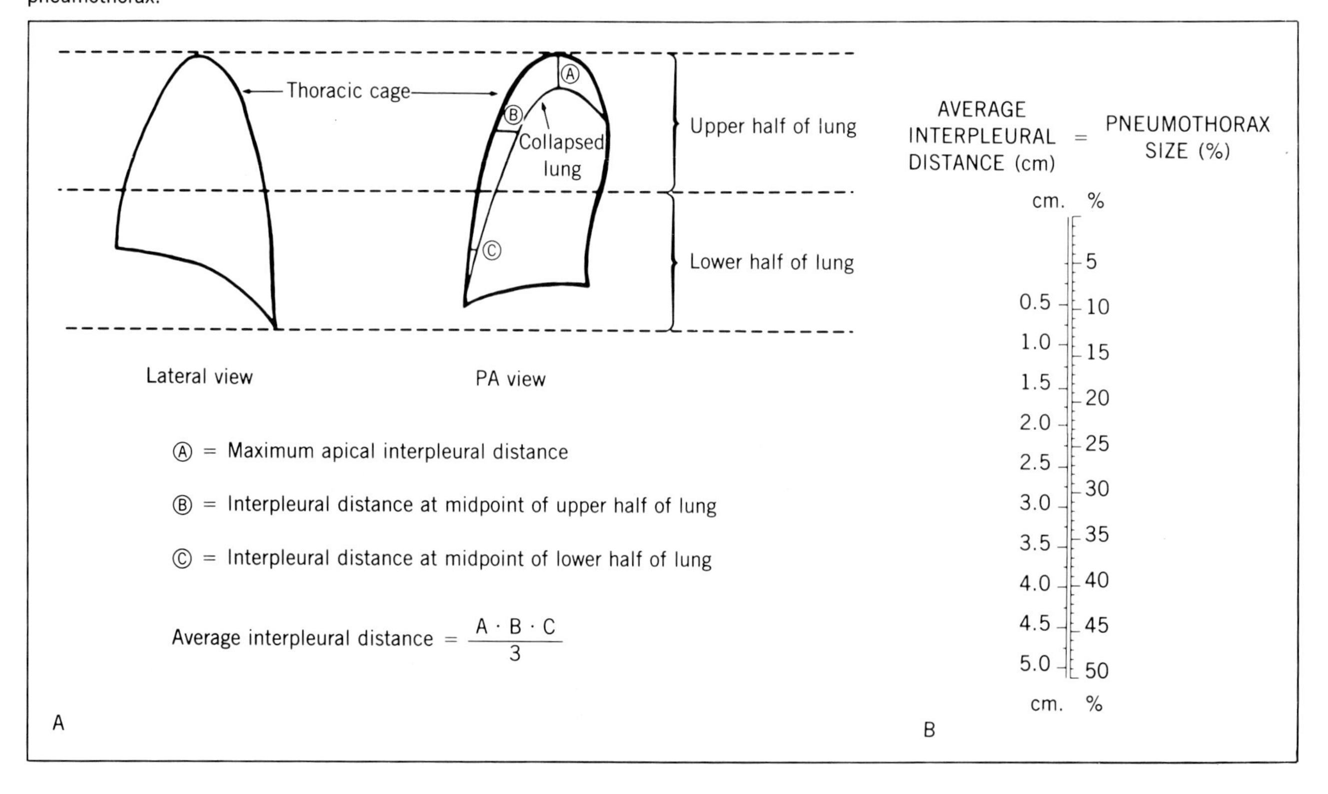

TABLE 5-1. Causes of Pneumothorax

Spontaneous
Traumatic
Iatrogenic following diagnostic or therapeutic procedures; barotrauma
Esophageal rupture
Pulmonary disease
Bleb, bulla, pneumatocele
Asthma
Cystic fibrosis
Eosinophilic granuloma
Scleroderma
Bronchopleural fistula
Pneumonoconiosis
Metastasis, especially osteosarcoma

ment expanded. The distribution of air is also influenced by patient position, as air rises to the nondependent portion of the thorax.

Subpulmonic pneumothorax is seen as a lucent area outlining the anterior costophrenic sulcus projected over the right or left upper quadrant [66] (Fig. 5-11) or as only a deep lateral costophrenic sulcus [67] (Fig. 5-12) on the involved side.

The causes of pneumothorax are listed in Table 5-1.

LINES OF PLEURAL REFLECTION

The recognition and analysis of the lines of pleural reflection allow one to recognize pathology and the location of the pathology within the mediastinum.

The anterior junction line is where the anterior aspects of the right and left lungs come together. Superiorly, it is seen as a *Y*-shaped structure (Fig. 5-13) to the left of the midline, never extending beyond the manubrium without diverging. Inferiorly, it never extends to the level of the diaphragm and is seen as an inverted *V*, separated by the heart. When the anterior junction line is normal, one can safely assume that there is no mass between the ascending aorta and the sternum. The anterior junction line may be displaced by an anterior mediastinal mass, adenopathy, hemorrhage, or herniation of one lung across the midline (Fig. 5-14).

The posterior junction line (Fig. 5-15) is seen when the

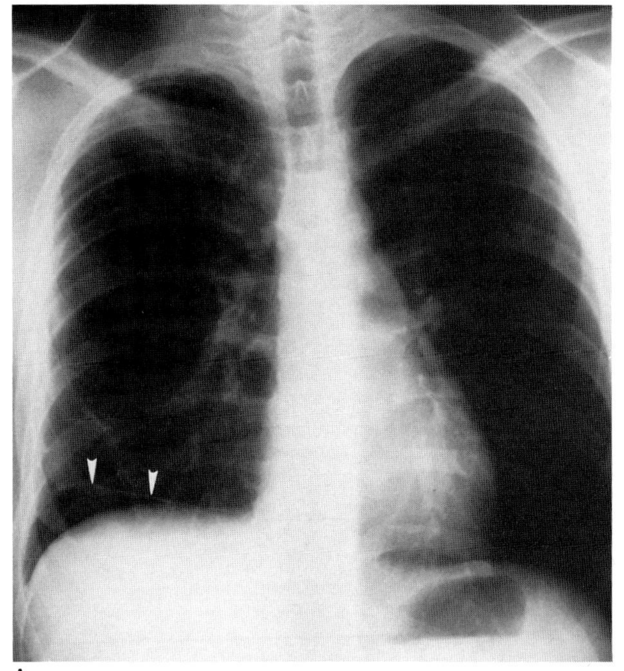

A

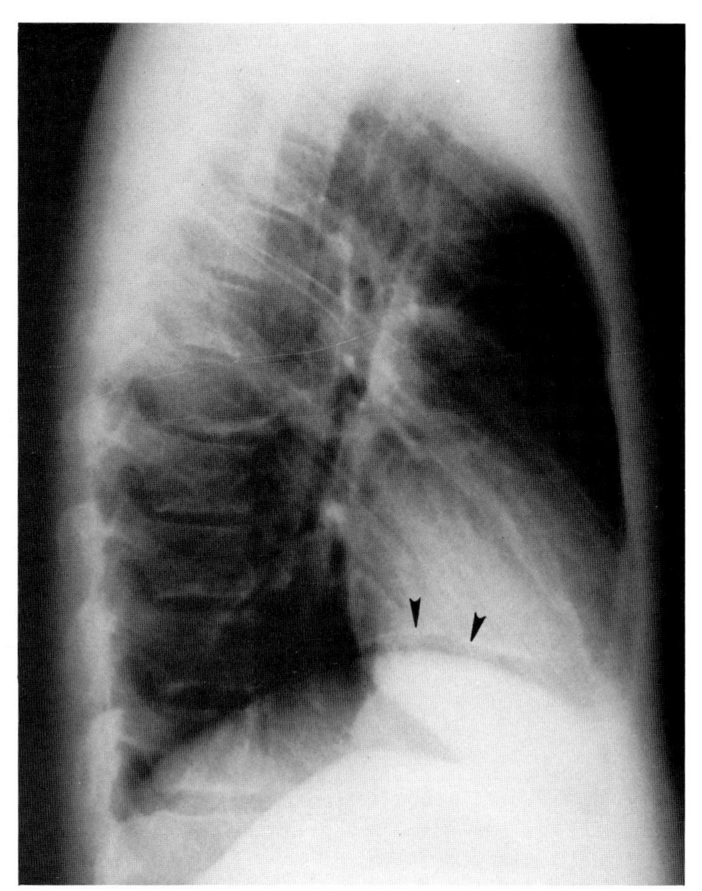

B

Fig. 5-11. *A.* Pleural line (arrowheads) in subpulmonic pneumothorax. *B.* Lateral view of the patient in *A.* Arrowheads point to the pleural line. *C.* Pleural line (arrows) in subpulmonic pneumothorax. *D.* Lateral view of the patient in *C.* Arrows point to the pleural line. *E.* Lucent air beneath the consolidated lung and the diaphragm (arrows) in subpulmonic pneumothorax.

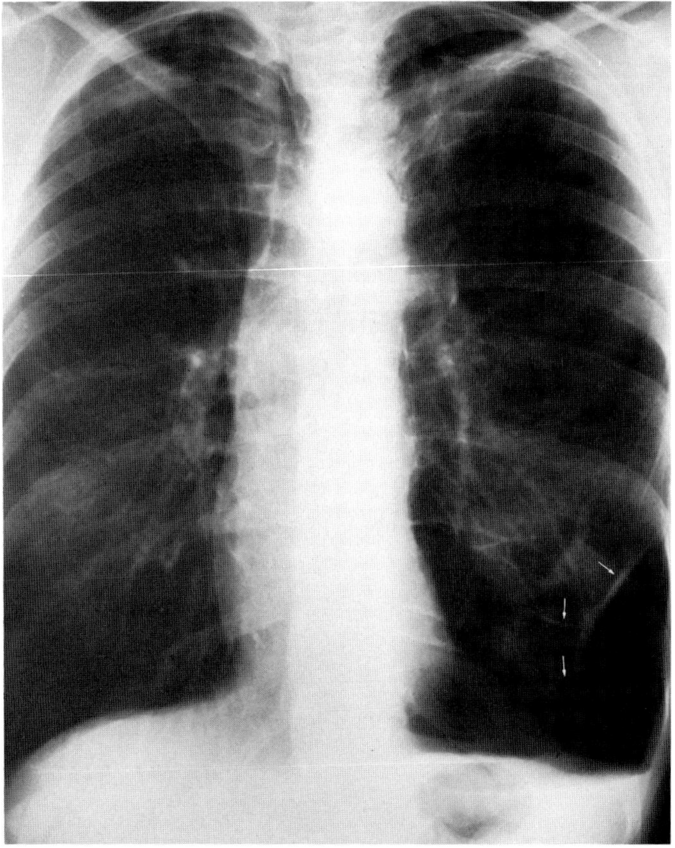

C

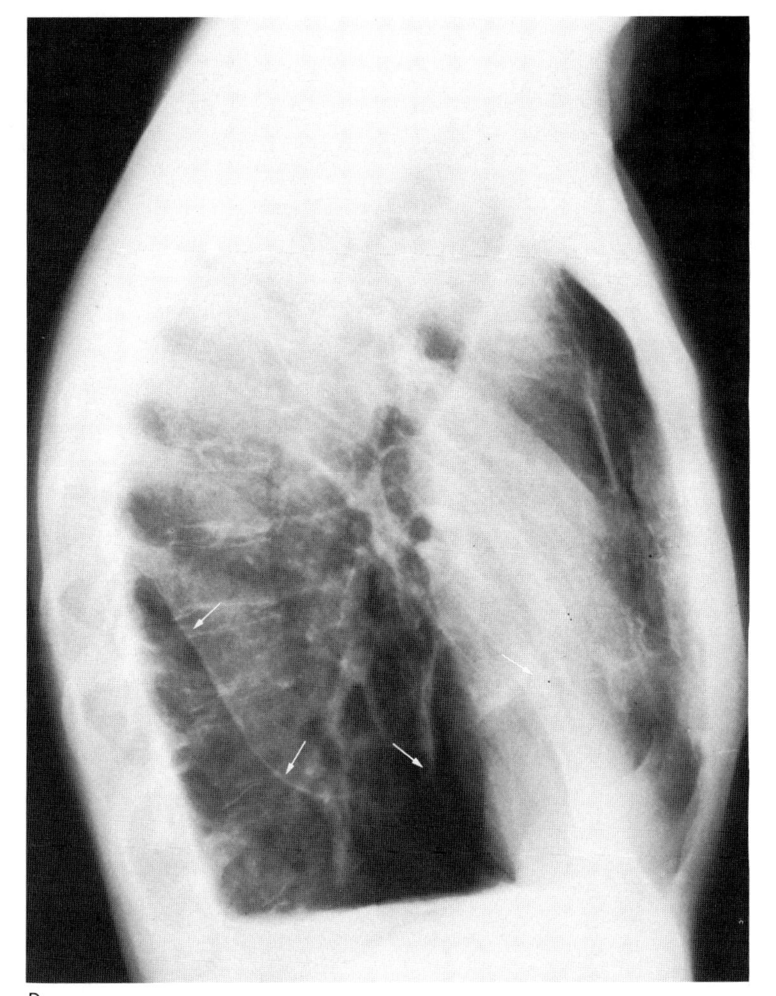

D

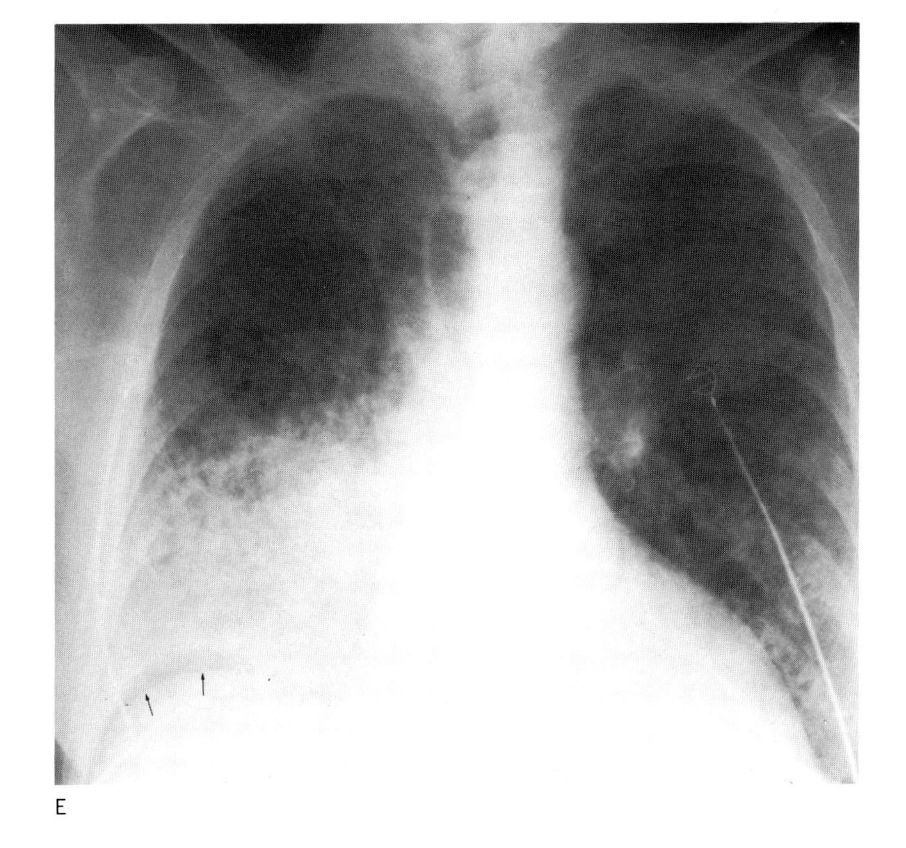

E

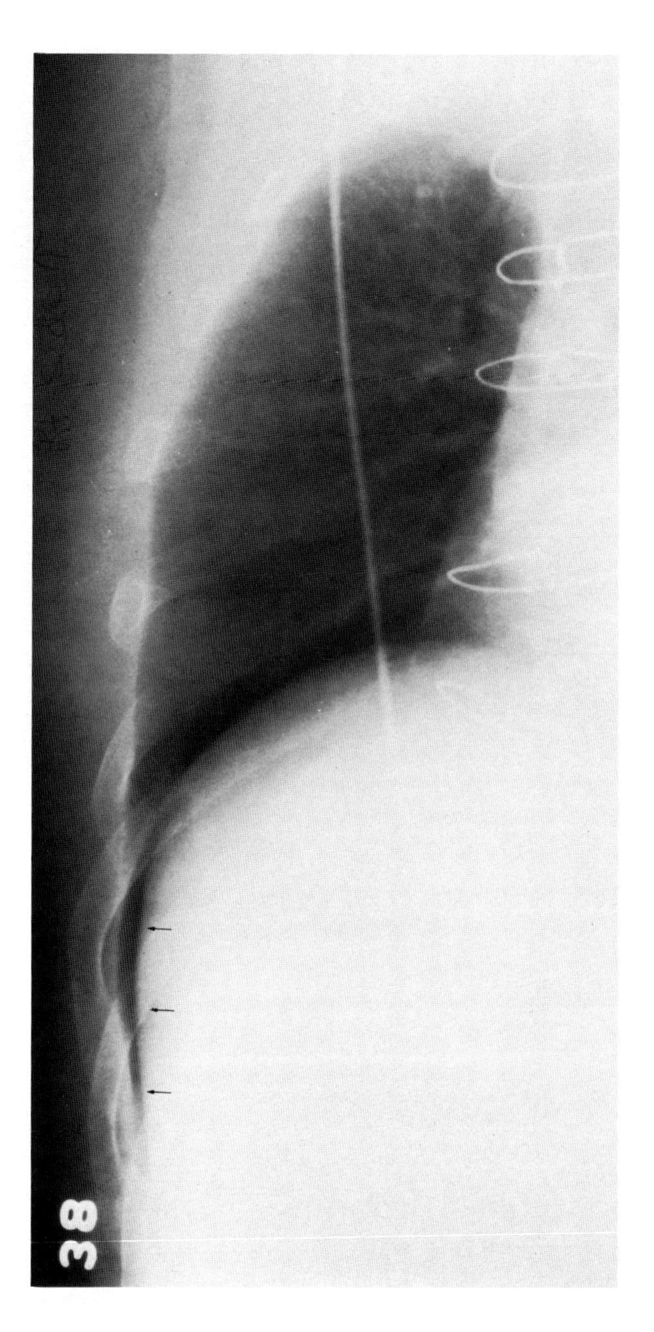

Fig. 5-12. Deep sulcus (arrows) produced by subpulmonic pneumothorax.

posterior aspects of the right and left lungs come in contact with each other. The width of the posterior junction line varies, depending on how close the lungs approximate each other. It is a straight line, in the midline, extending above the manubrium, superiorly, and into the diaphragm, inferiorly. It is displaced by posterior mediastinal masses, esophageal dilatation, aortic arch anomaly, hemorrhage, and herniation of lung.

The right paraesophageal line is seen as a line concave to the right at its upper third up to the level of the right bronchus, where it is pulled laterally by the azygous vein. It then continues obliquely inferiorly to the level of the esophageal

Fig. 5-13. The anterior junction line (arrows) deviating laterally on both sides below the level of the clavicle.

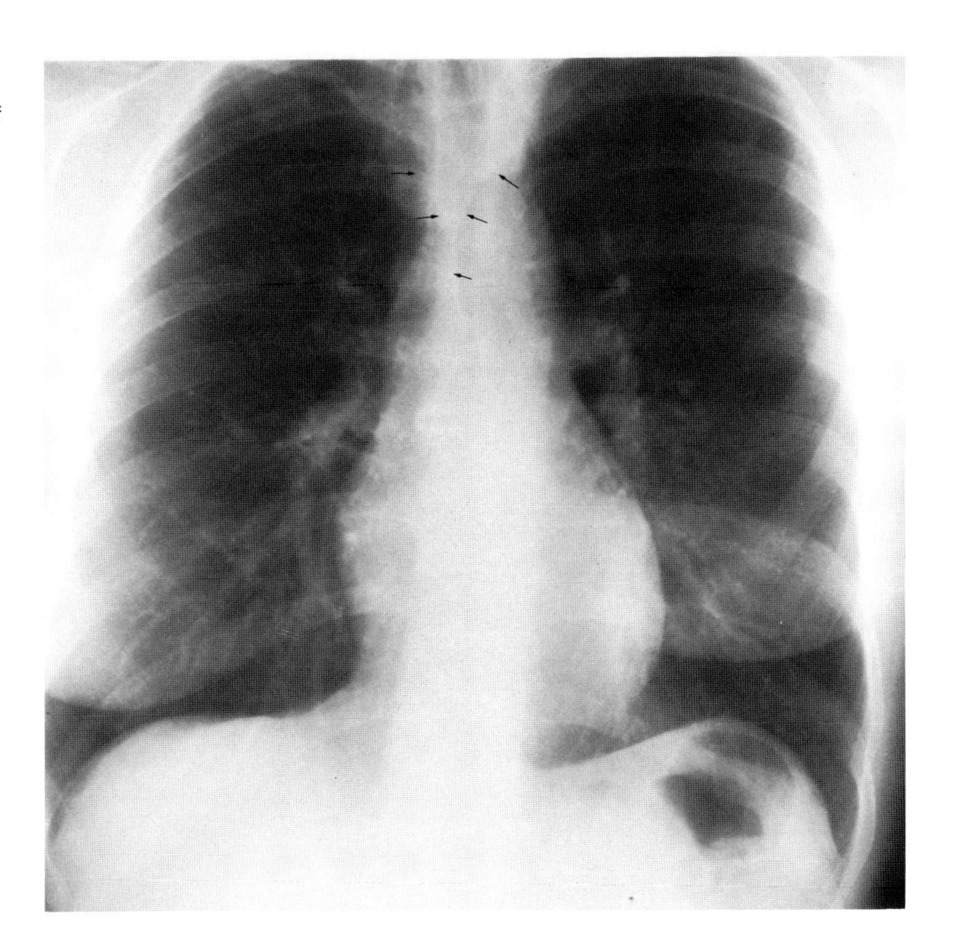

Fig. 5-14. Displaced anterior junction line (arrows). Displacement was produced by lung herniating across the midline. Note that the heart is also displaced to the right.

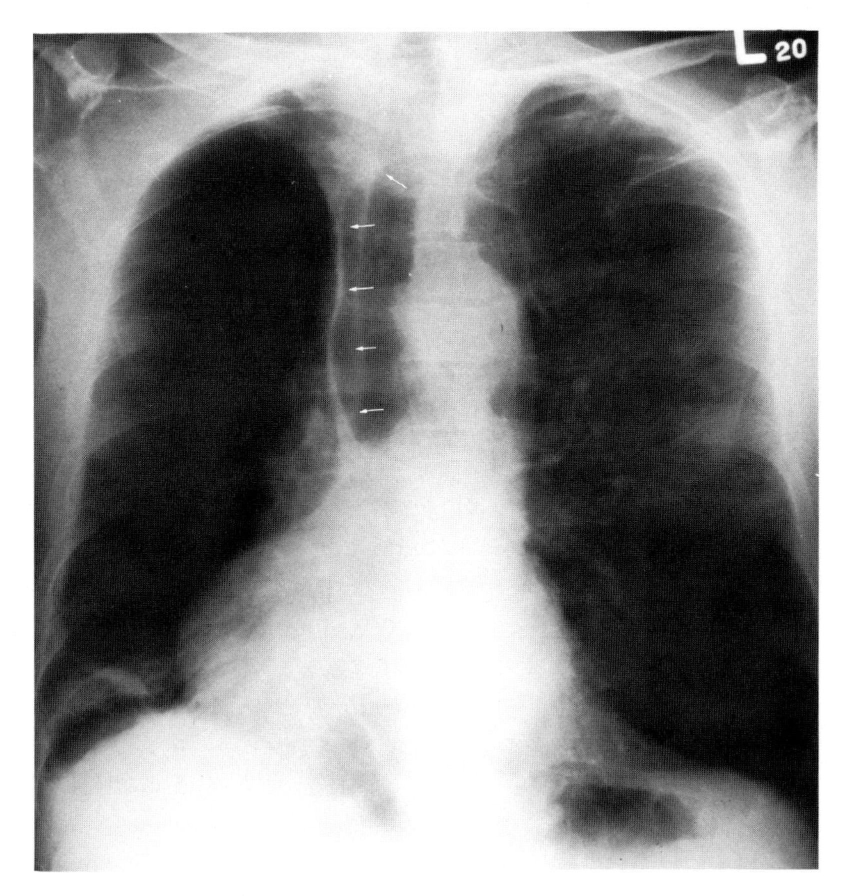

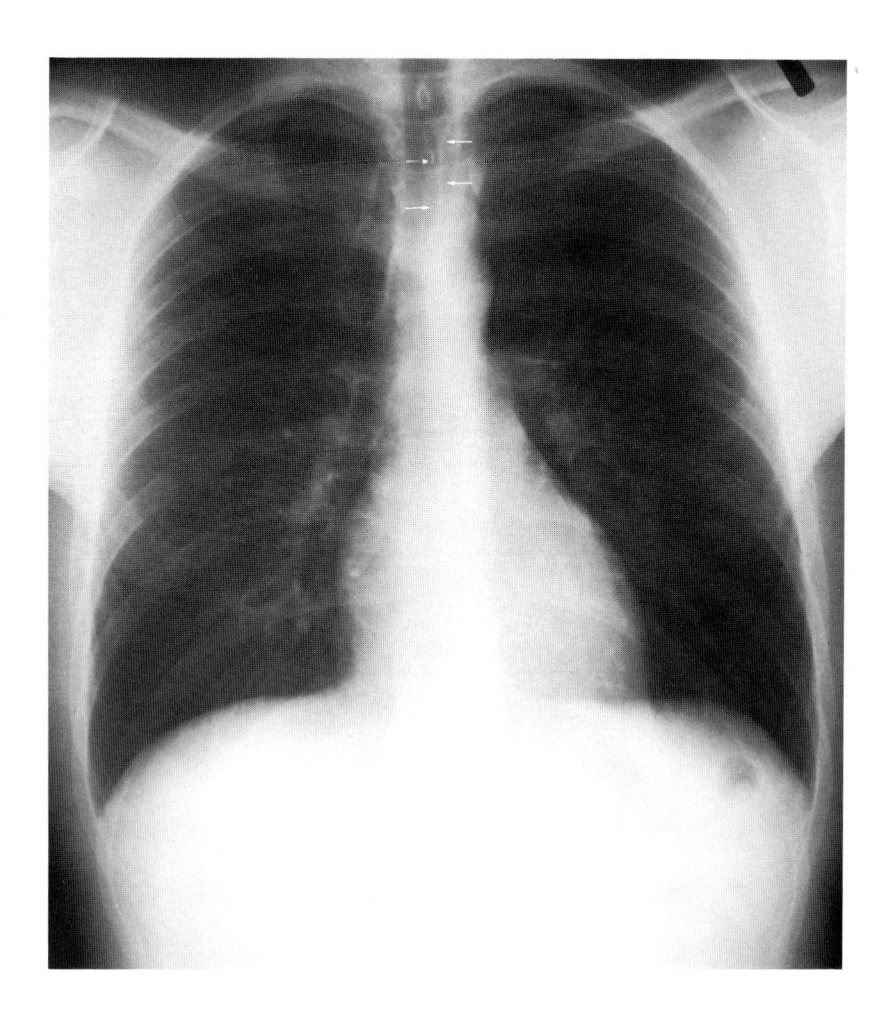

Fig. 5-15. White line between arrows represents the posterior junction line.

hiatus. Esophageal lesions and paraesophageal modal enlargements can displace this line (Fig. 5-16).

The paraspinal lines parallel the spine on both sides. The right is more consistently seen than the left. On the left, the paraspinal line must be distinguished from the descending aorta. It is seen as a line edged in white, whereas the aorta is seen as a line edged in black [68,69]. Paraspinal lesions such as a tuberculous abscess, extramedullary hematopoiesis, neurogenic tumors such as neuroblastoma, and hematomas can displace this line (Fig. 5-17).

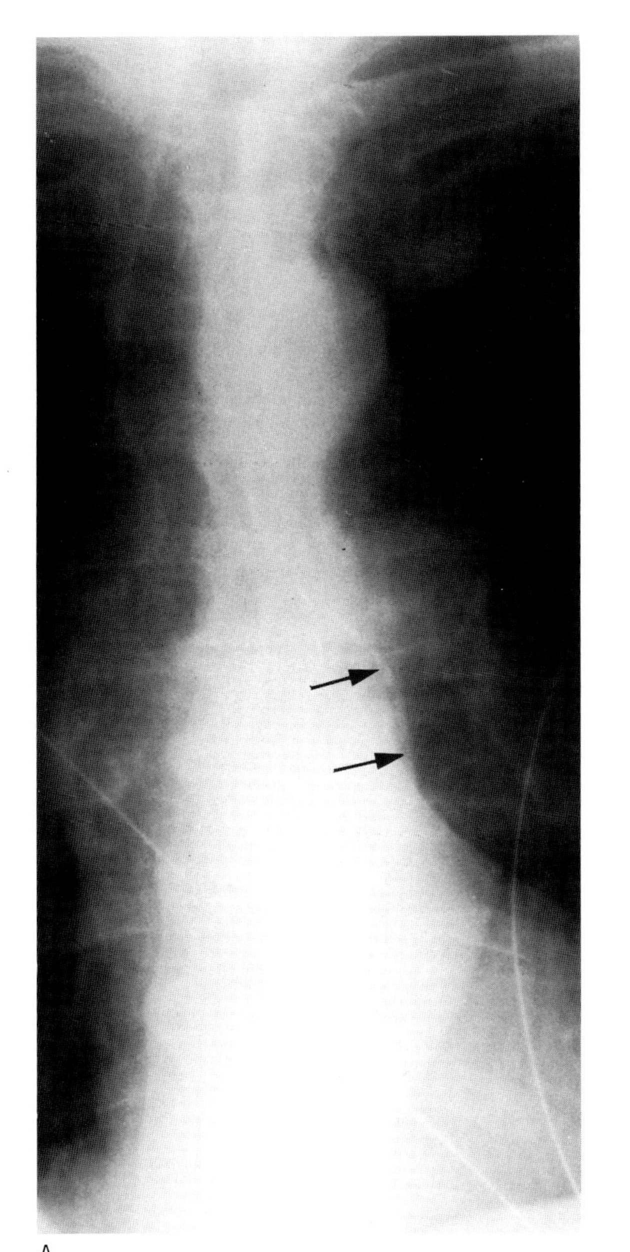

A

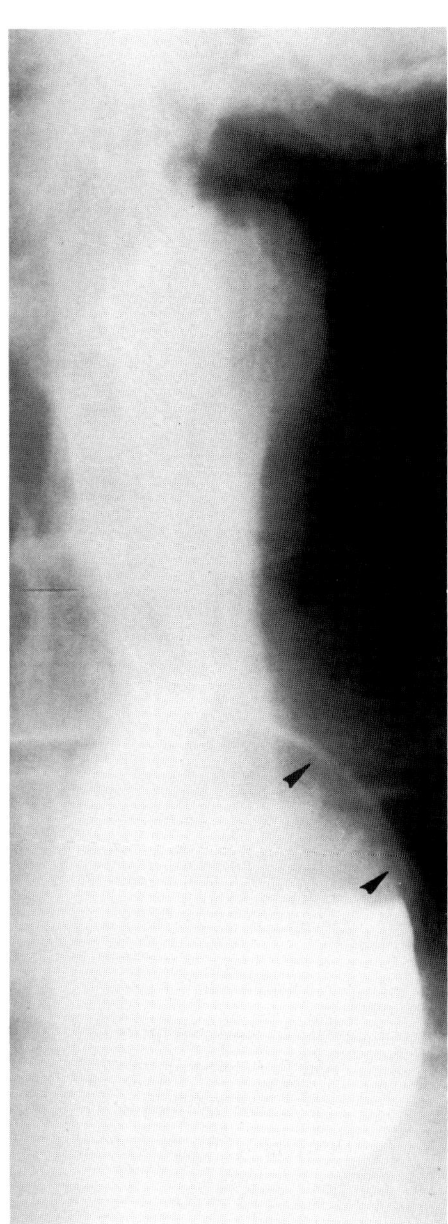

B

Fig. 5-16. *A.* A displaced paraesophageal line (arrows). *B.* Large barium-filled esophageal diverticulum (arrowheads) that produced the displacement.

Fig. 5-17. A laterally displaced left paraspinal line (arrows) secondary to a posterior mediastinal mass. Arrowheads point to the descending aorta.

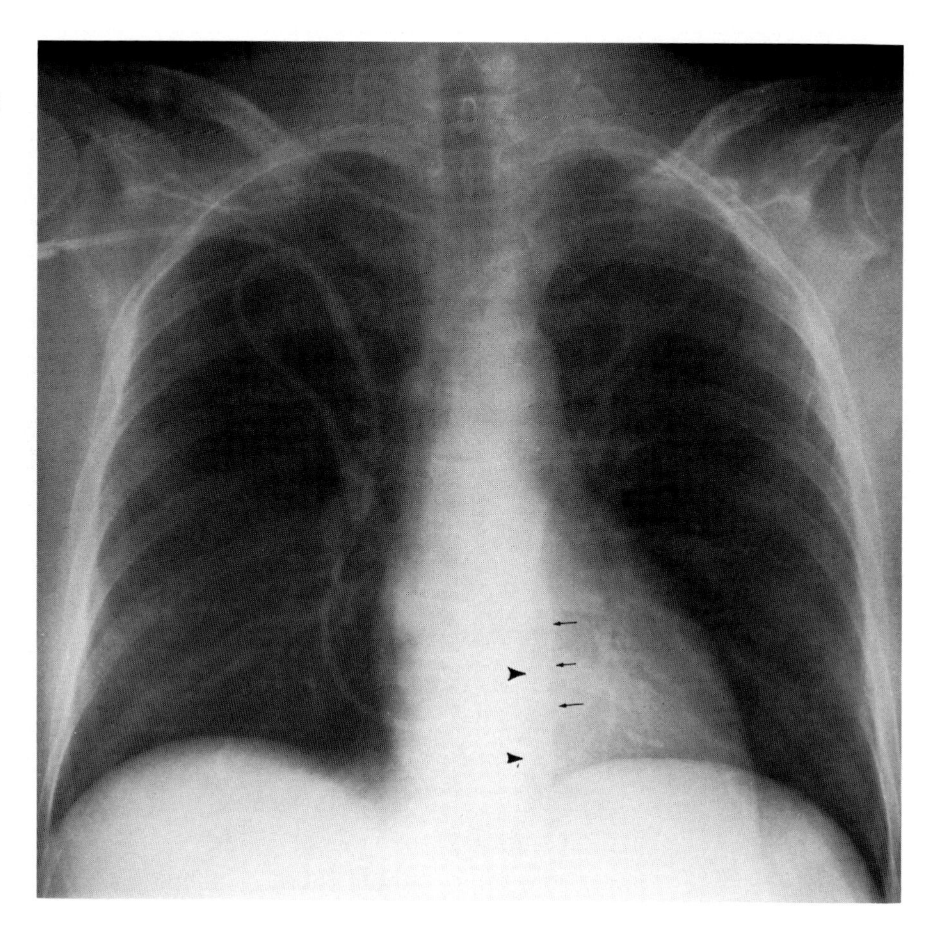

STEP 6

EVALUATE THE POSITION OF TUBES AND CATHETERS

The position of tubes and catheters in a patient, and possible complications arising from them, should be assessed as soon as they have been inserted.

ENDOTRACHEAL TUBE

To evaluate endotracheal tube position, one must take into consideration the neck position using the mandible as a guide [70]. When the neck is extended (the inferior border of the mandible at or above C4), the tip of the endotracheal tube should be 7 ± 2 cm from the carina. With the neck in a neutral position (the inferior border of the mandible at C5-6 level), the endotracheal tube tip should be 5 ± 2 cm from the carina. With the neck in flexion (the inferior border of the mandible at T1 or below), the endotracheal tube tip should be 3 ± 2 cm from the carina. In addition to evaluating the position of the tube, one must evaluate the size of the tube and the cuff. Ideally, the tube should be one-half to two-thirds the width of the trachea. The cuff should be inflated only to the tracheal diameter and should not cause the walls to bulge (Fig. 6-1).

TRACHEOSTOMY TUBE

Tracheostomy tubes, unlike endotracheal tubes, do not move with flexion and extension. The tip of a tracheostomy tube should be one-half to two-thirds of the way between the stoma and the carina. Its inner diameter should be two-thirds of the tracheal lumen, its long axis should be parallel to the tracheal wall, and its distal end should not abut any portion of the tracheal wall (Fig. 6-2).

Fig. 6-1. Diameter of endotracheal tube cuff (black arrows) is much greater than tracheal diameter (arrowheads). Tip of endotracheal tube (open arrow) is too low; it is at the carina directed toward the right bronchus.

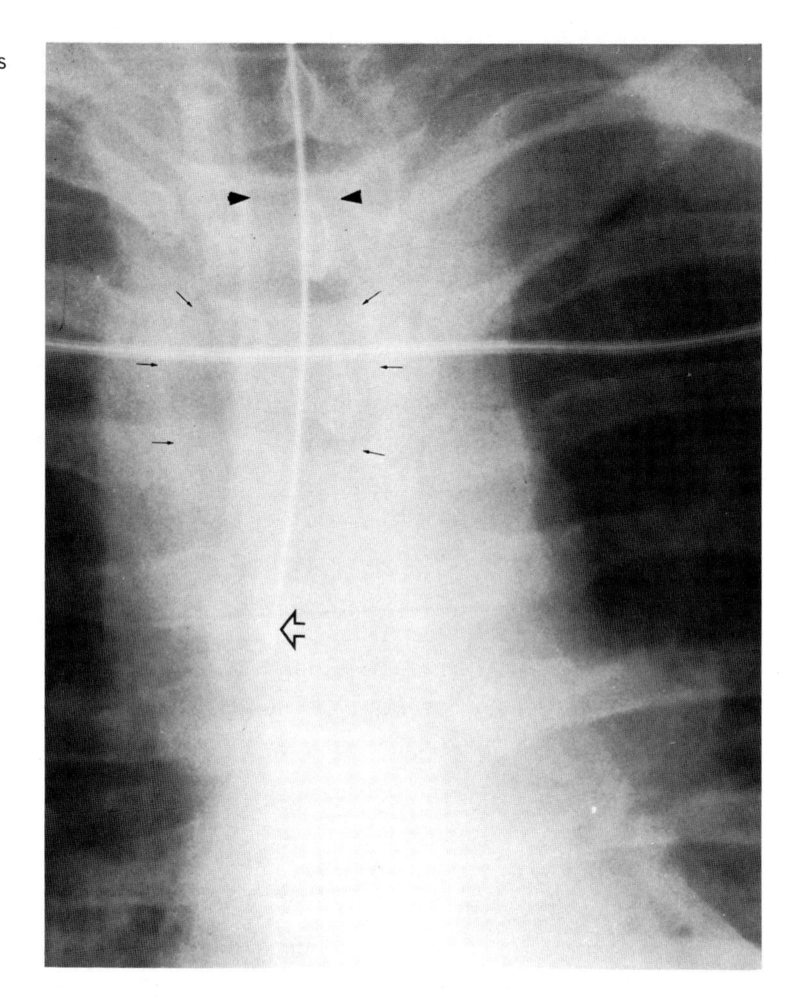

Fig. 6-2. An overdistended tracheostomy tube cuff (arrow). Arrowheads demonstrate the normal tracheal diameter. Open arrowhead points to tip that abuts the left lateral tracheal wall.

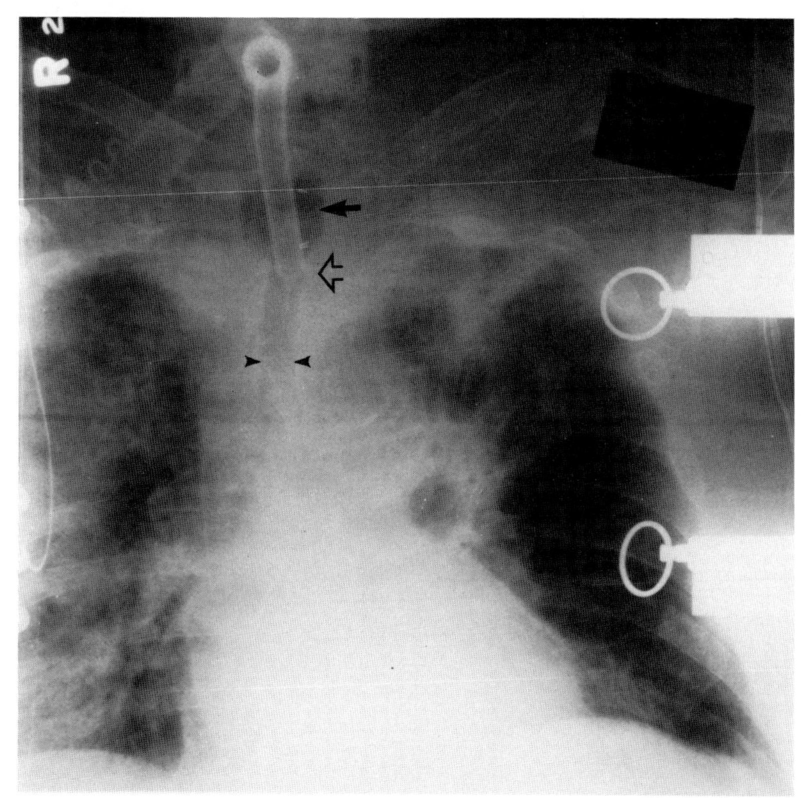

CENTRAL VENOUS LINE

Central venous lines should be seen in the superior vena cava between the right internal jugular vein and subclavian vein junction (approximately at the level of the first anterior rib) and the right atrium (Fig. 6-3).

SWAN-GANZ CATHETER

The tip of a Swan-Ganz catheter should be in the right or left pulmonary artery or in the proximal portion of a lobar branch, up to approximately 8.5 cm from the main pulmonary artery bifurcation (Fig. 6-4). Its tip should be below the level of the left atrium in the lateral view.

INTRAAORTIC COUNTERPULSATION BALLOON

An intraaortic counterpulsation balloon should be at least 4 cm below the upper part of the aortic knob (Fig. 6-5).

CARDIAC PACEMAKER

A transvenous cardiac pacemaker's tip should be in the apex of the right ventricle (Fig. 6-6A); if a bipolar type is used, the other end should be in the right atrium (Fig. 6-6B). The ventricular tip should be directed anteriorly (Fig. 6-7), 3 to 4 mm beneath the epicardial fat stripe [71]. The pulse generator should be free of air to avoid a system malfunction [72]. This should be watched for in patients with subcutaneous emphysema.

CHEST TUBE

Chest tubes for pleural drainage should have their end and side holes located intrapleurally. The optimal position for draining a pneumothorax is anterosuperior at about the level of the third intercostal space, axillary line, and posteroinferior at about the eighth intercostal space, mid to posterior axillary line for hydro- or hemothorax.

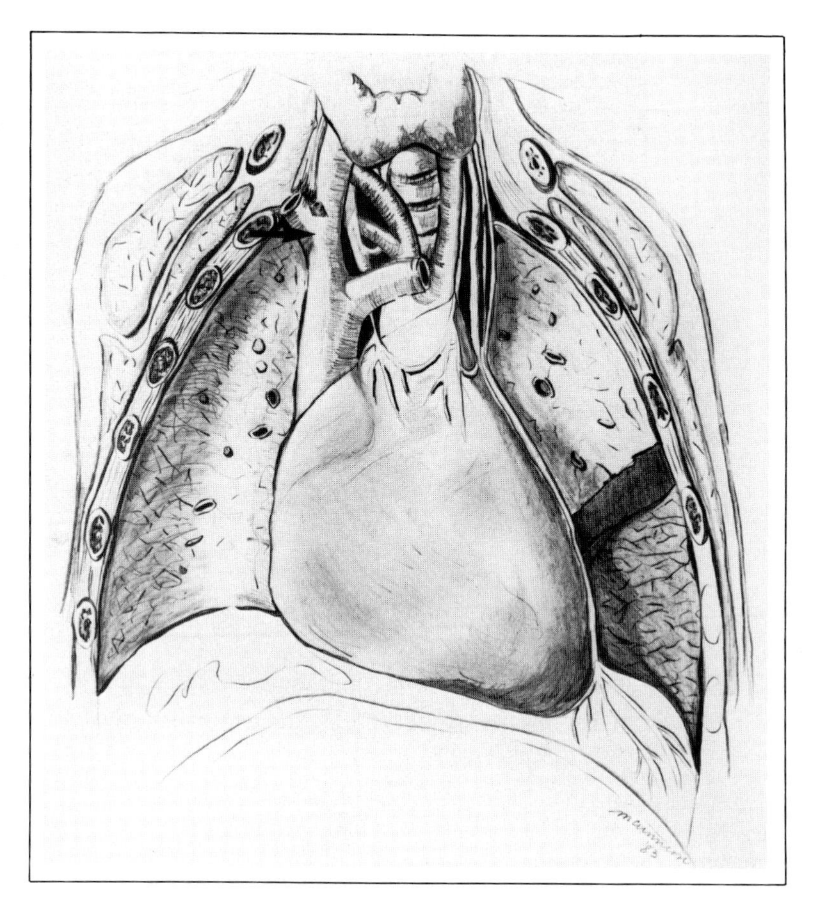

Fig. 6-3. Junction of the right brachiocephalic vein and internal jugular vein (arrow). Central venous pressure–line tip should be beyond this point.

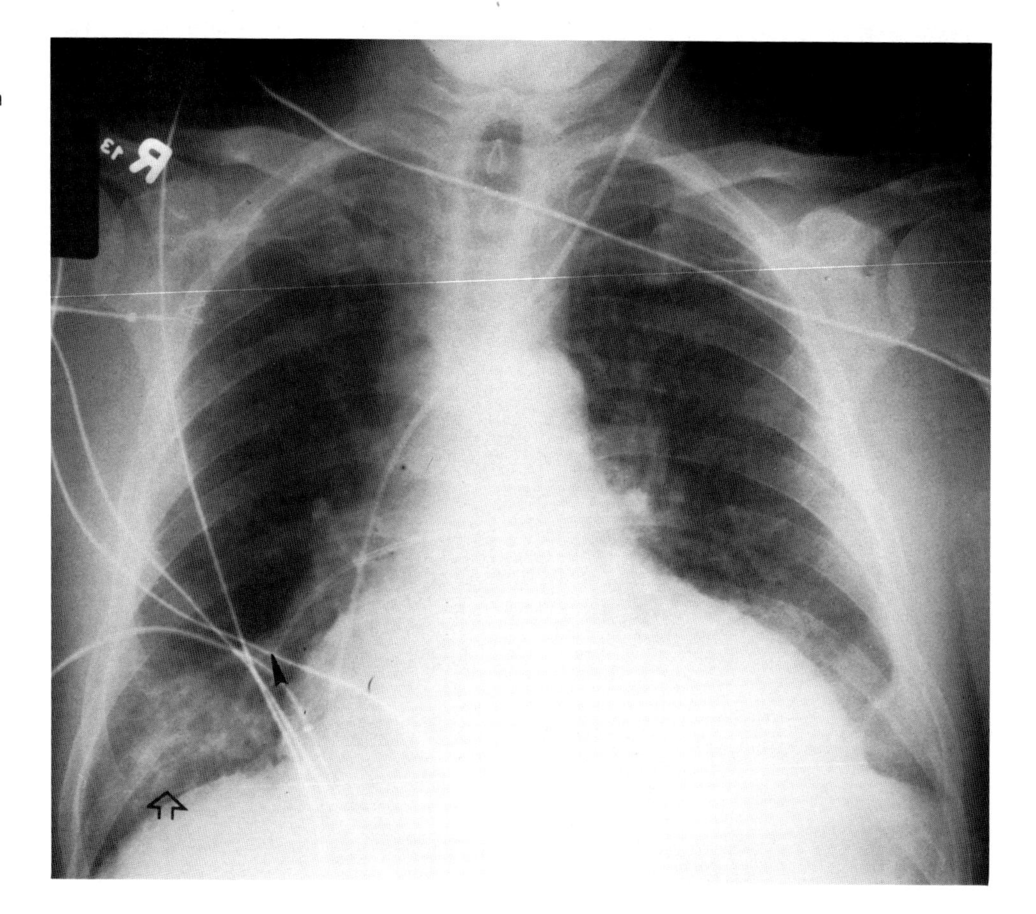

Fig. 6-4. Tip of a Swan-Ganz catheter (arrowhead) that is located in a very distal position and a resulting pulmonary infarct (open arrow).

Fig. 6-5. Tip of an intraaortic counterpulsation balloon (arrowhead) that is too proximally located in the aortic arch and can occlude the left carotid and subclavian arteries.

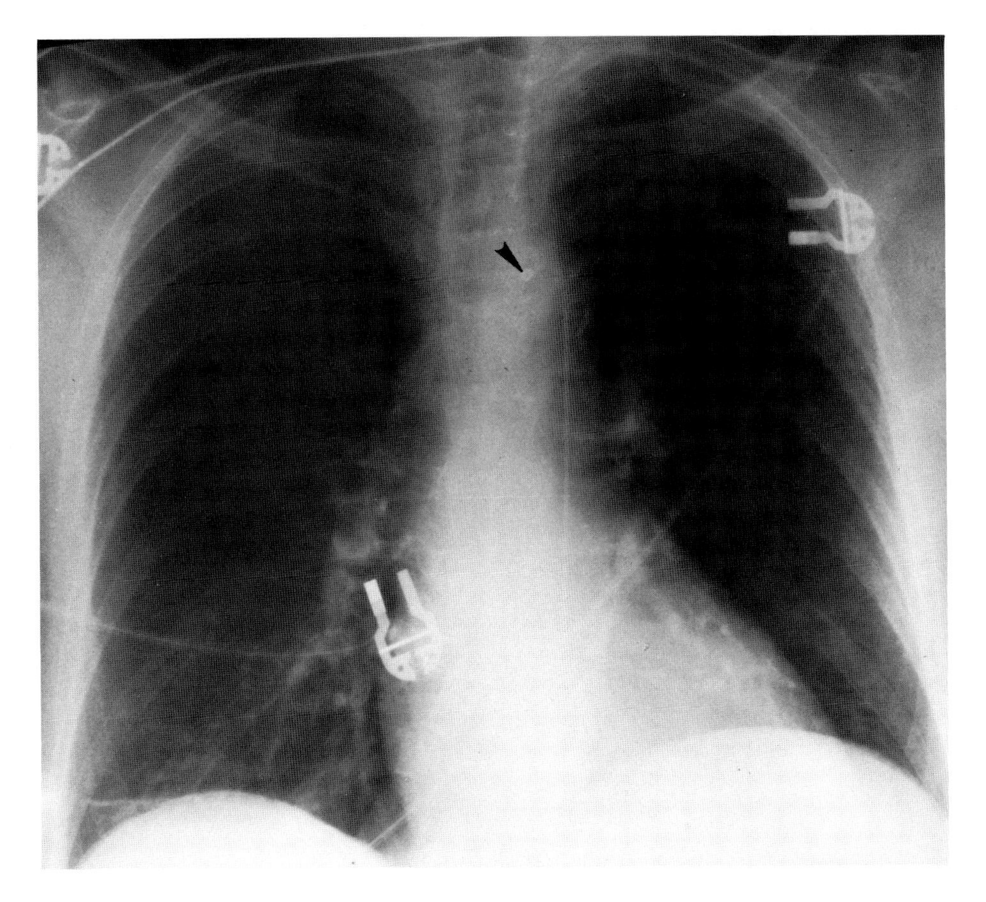

Fig. 6-6. *A.* Tip of a pacemaker (arrowhead) in the right ventricle. *B.* Atrial tip of a bipolar pacemaker (arrowhead).

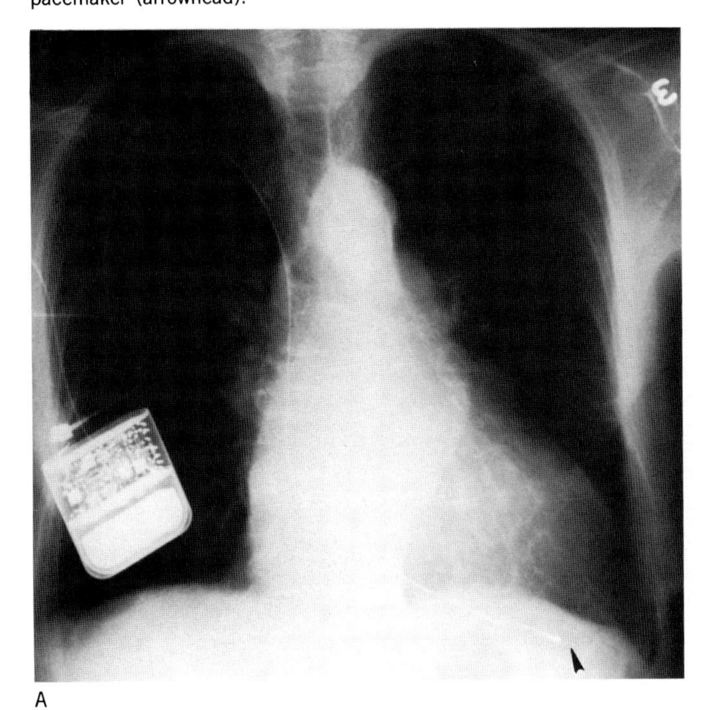

A

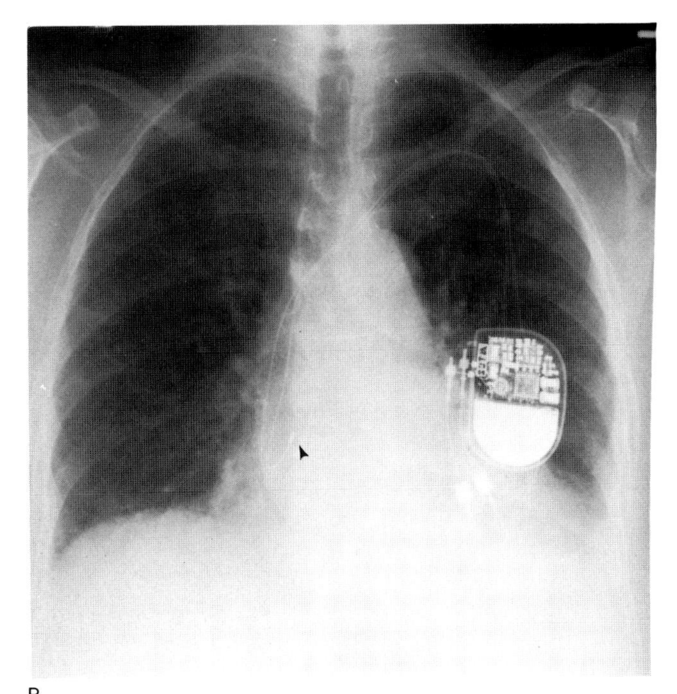

B

Fig. 6-7. Arrows point to a pacer within the coronary sinus. Note posterior direction taken by the pacer wire in *B*. Double arrows are at the tip of the pacer wire. (Courtesy of Dr. Lawrence J. Moss.)

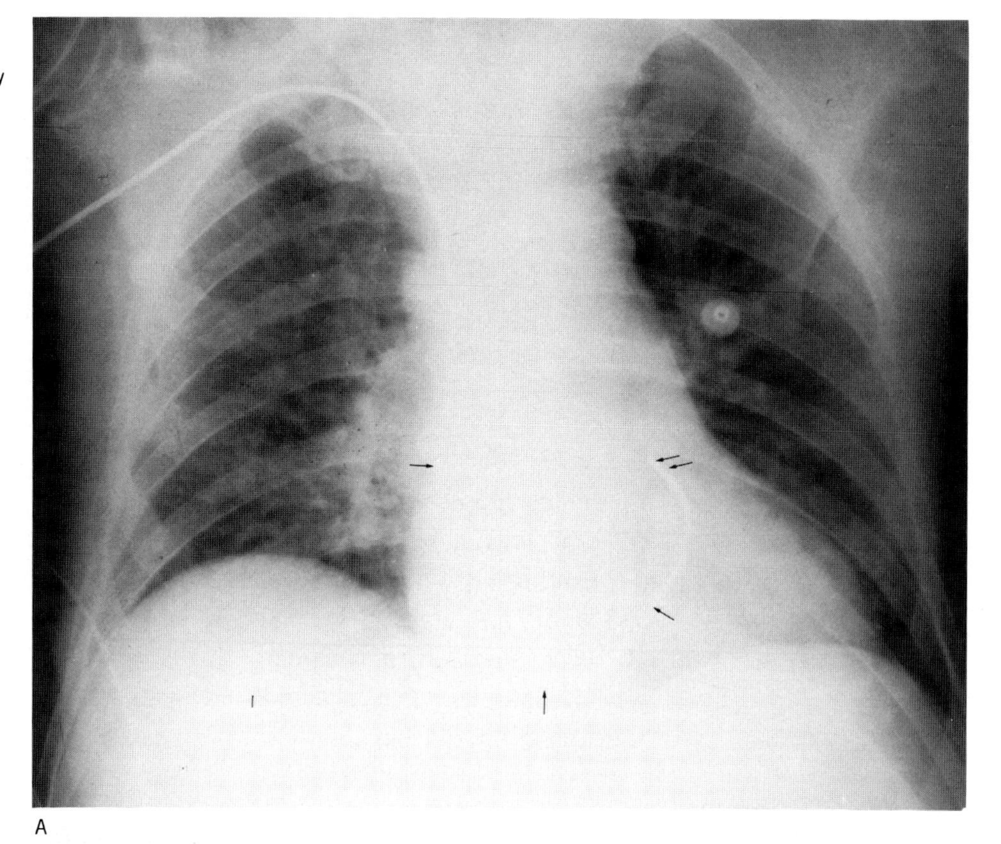

A

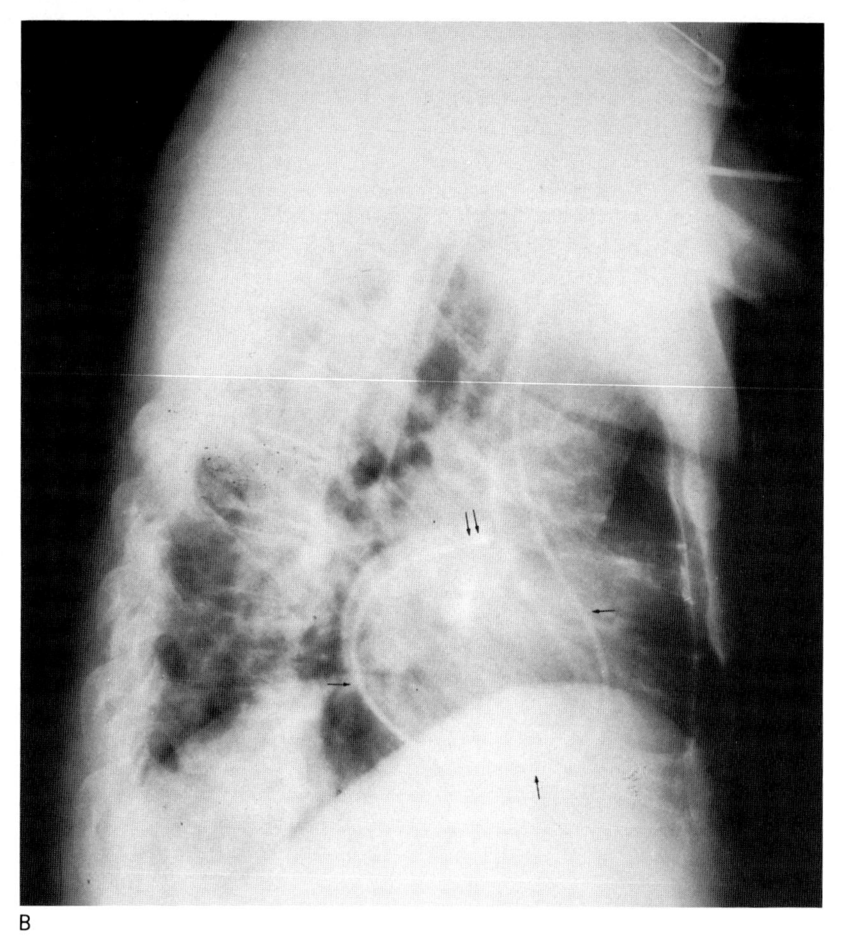

B

STEP 7

EVALUATE SOFT TISSUES, BONY STRUCTURES, AND VISIBLE PORTIONS OF THE UPPER ABDOMEN

The bony structures and chest wall soft tissues, as well as the visible parts of the upper abdomen, should be checked for pathology. Each structure (ribs, clavicles, scapulae [Fig. 7-1], humeri [Fig. 7-2], vertebrae, breasts, chest wall [Figs. 7-3 and 7-4A and B], subdiaphragmatic area [Fig. 7-5A and B], liver, stomach, and spleen) should be assessed individually. If pathology is suspected in any area, specific film of the region should be obtained.

Fig. 7-1. *A*. Arrows point to a denser coracoid process of the right scapula as compared to the left (arrowheads). Coned-down view (*B*) of the scapula reveals blastic metastasis (arrows). (Courtesy of Dr. Ashley Davidoff.)

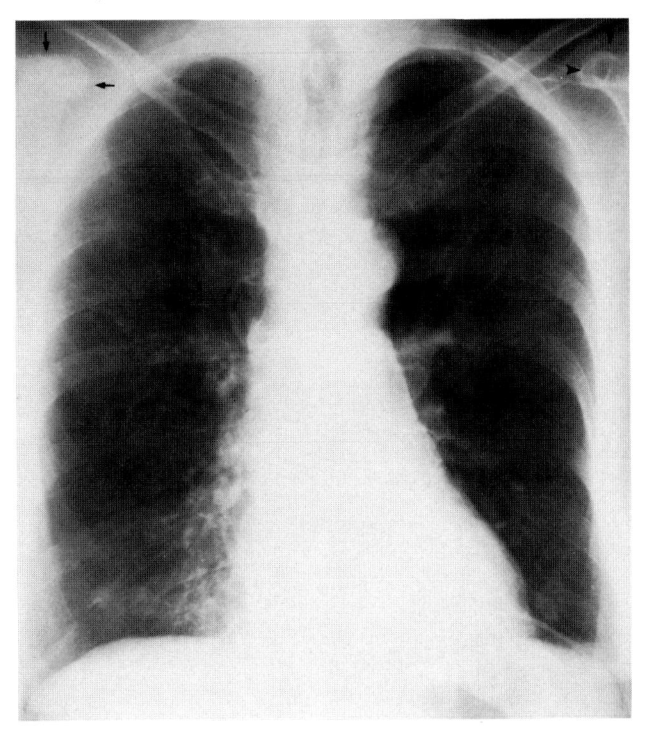

A

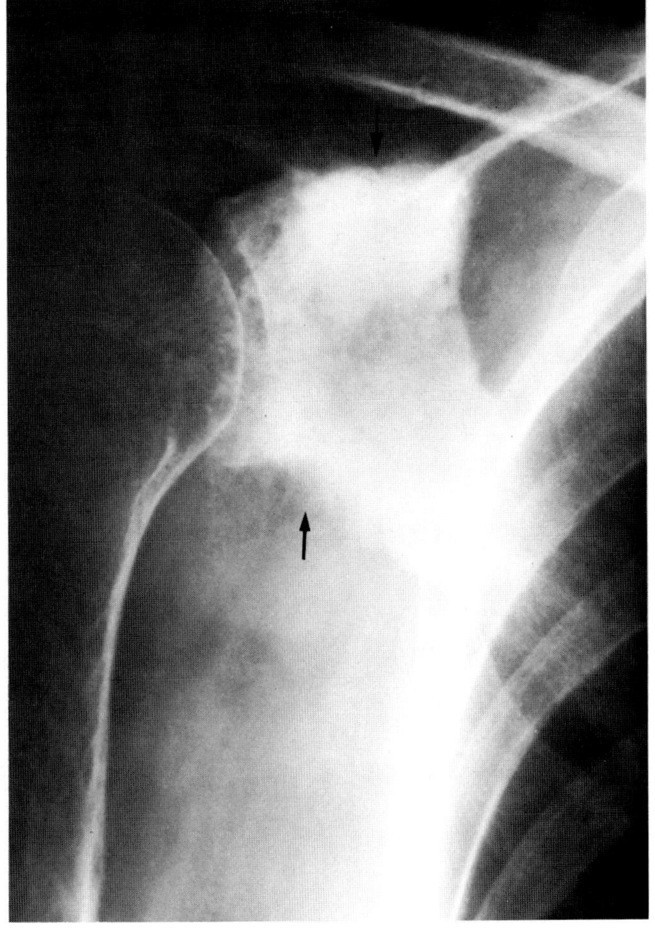

B

Fig. 7-2. *A.* Dislocated right shoulder (arrowhead). *B.* A close-up view of the dislocated shoulder. Note bone island in medial aspect (arrow). (Courtesy of Dr. Ashley Davidoff.)

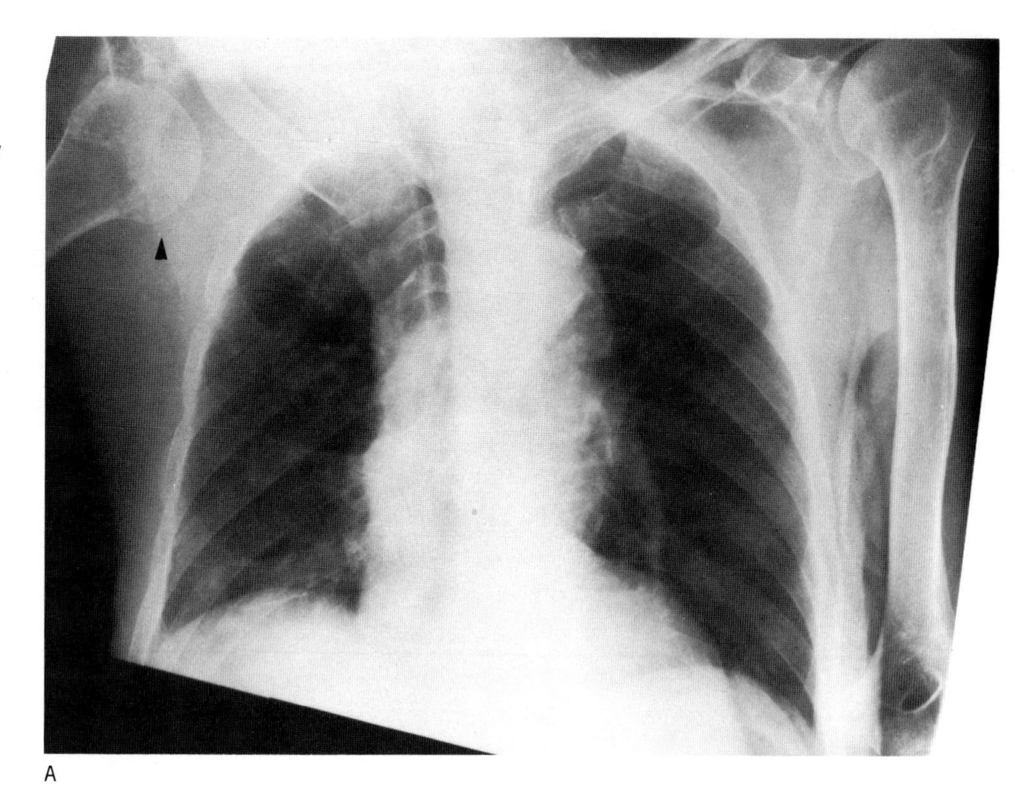

A

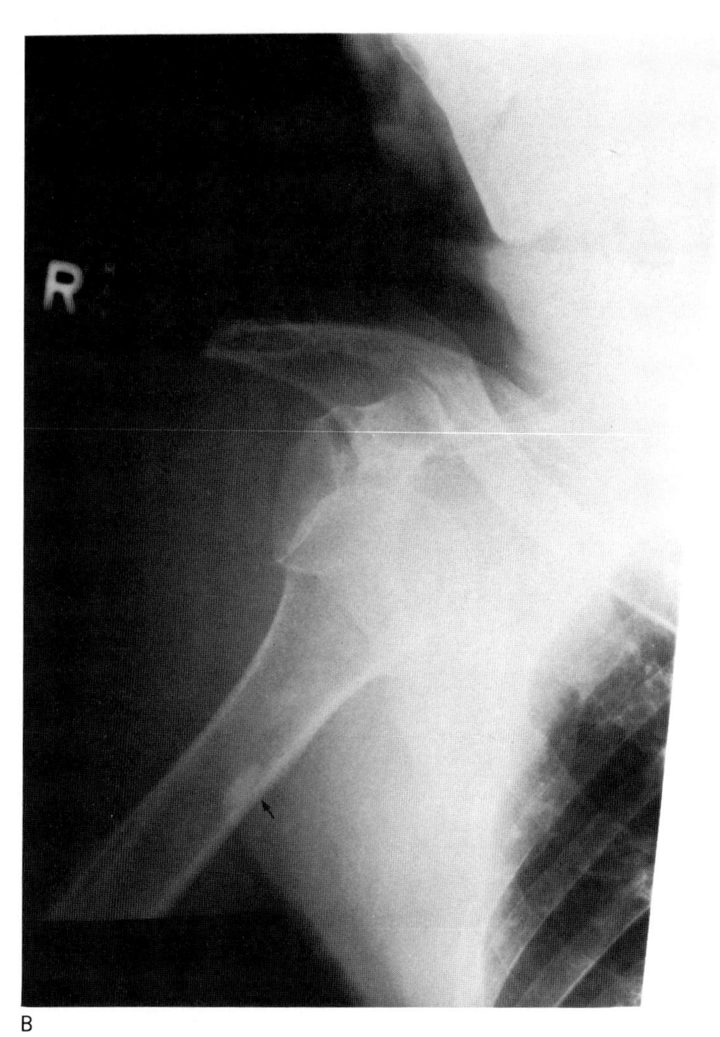

B

Fig. 7-3. Air in the soft tissues outlines the fasciculi of pectoral muscles (arrows) and air in the subcutaneous tissues of the supraclavicular area (arrowheads) in a patient with subcutaneous emphysema.

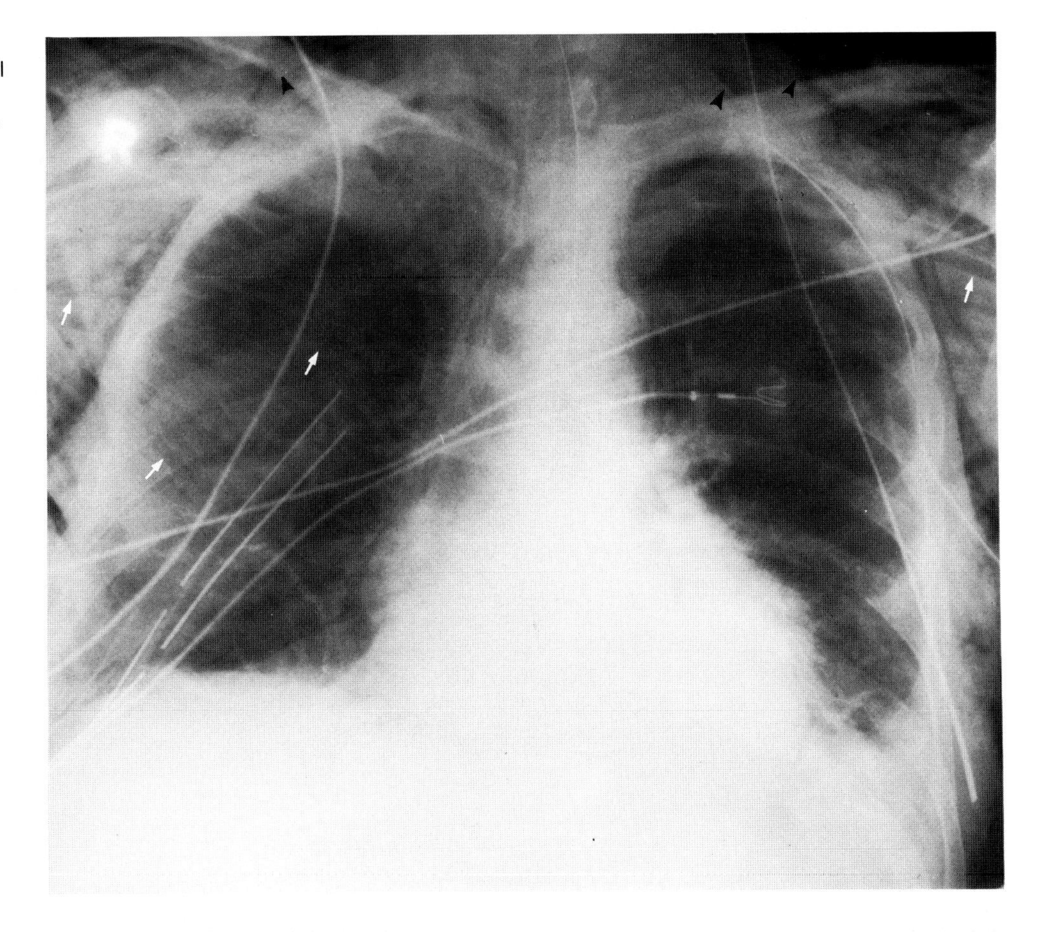

Fig. 7-4. *A.* Multiple nodular densities in a patient with neurofibromatosis. Note the nodules in the soft tissues of the shoulders. *B.* A few skin nodules (arrows) in a patient with neurofibromatosis.

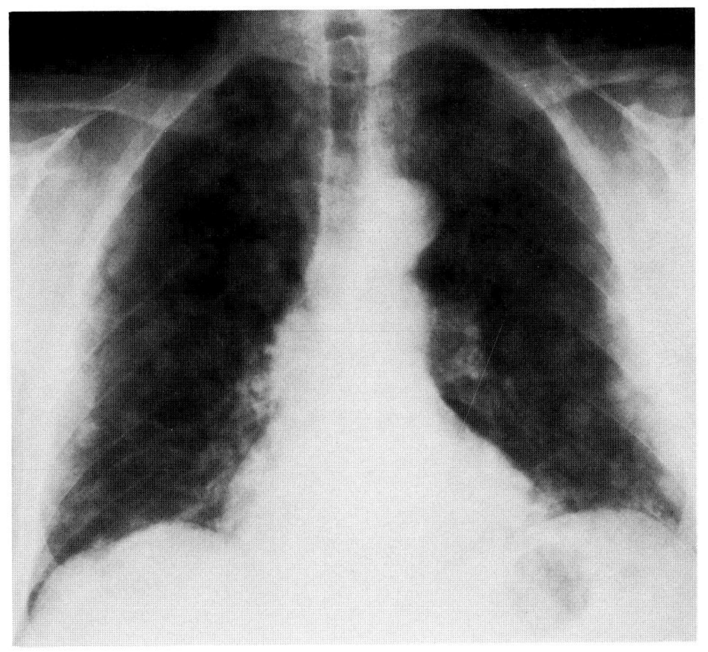

A

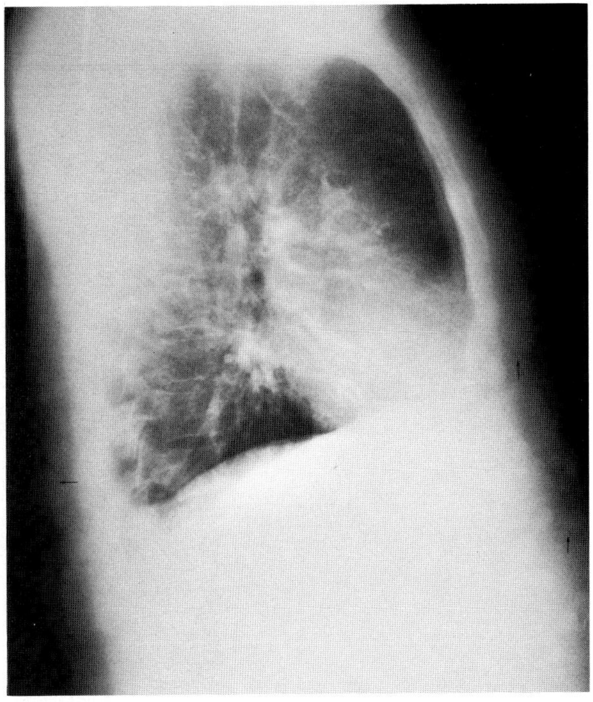

B

Fig. 7-5. *A.* Arrow over the liver points to large amount of free air underneath the diaphragm (arrowhead). *B.* Arrows point to small amounts of free air beneath the right and left hemidiaphragms.

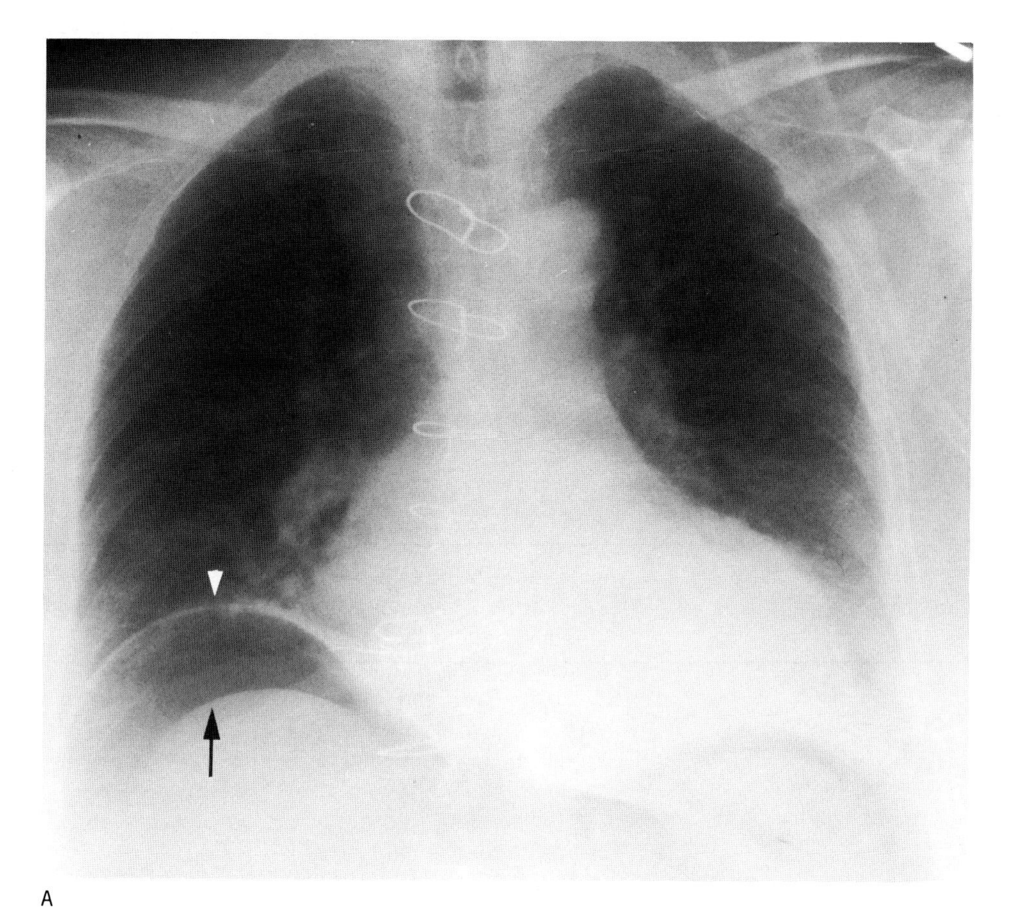

A

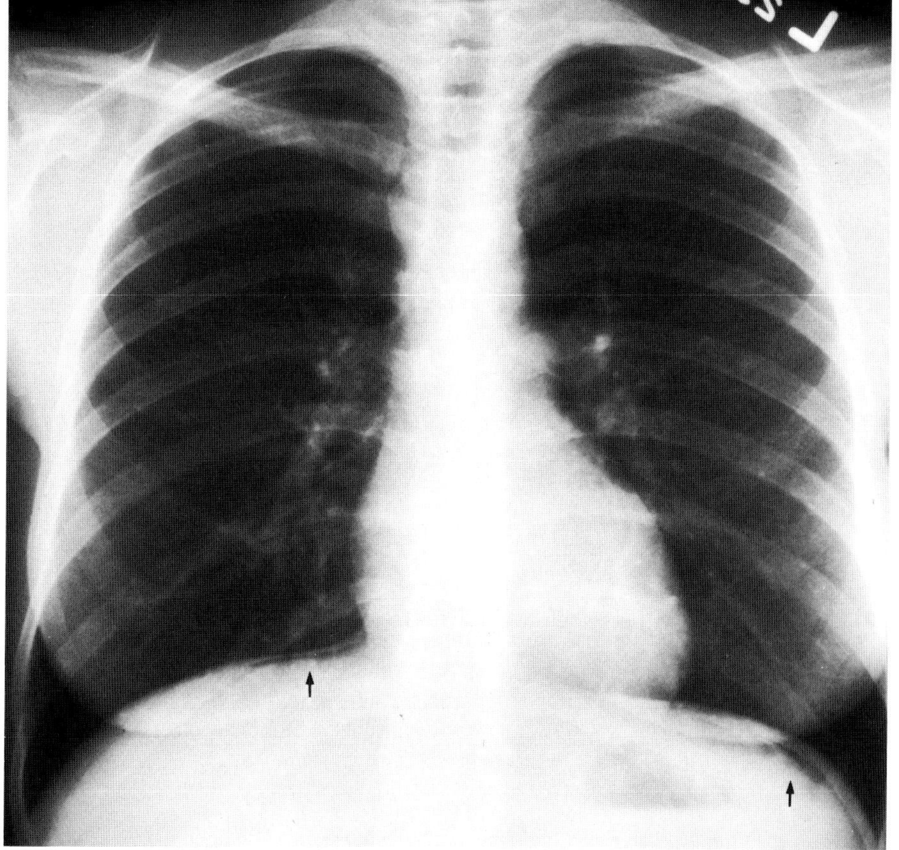

B

REFERENCES

1. Felson, B., and Felson, H. Localization of intrathoracic lesions by means of the postero-anterior roentgenogram: The silhouette sign. *Radiology* 55:363–374, 1950.
2. Felson, B. More chest roentgen signs and how to teach them. Annual oration in memory of L. Henry Garland, M.D., 1903–1966. *Radiology* 90:429–441, 1968.
3. Spratt, Jr., J.S., Ter-Pogossian, M., and Long, R.T.L. The detection and growth of intrathoracic neoplasms: The lower limits of radiographic distinction, of the antemortem size, the duration, and the pattern of growth as determined by direct mensuration of tumor diameters from random thoracic roentgenograms. *Arch. Surg.* 86:283–288, 1963.
4. Greening, R.R., and Pendergrass, E.P. Postmortem roentgenography with particular emphasis upon the lung. *Radiology* 62:720–725, 1954.
5. Newell, R.R., and Garneau, R. The threshold visibility of pulmonary shadows. *Radiology* 56:409–415, 1951.
6. Resink, J.E.J. Is a roentgenogram of fine structures a summation image or a real picture? *Acta Radiol.* 32:391–403, 1949.
7. Felson, B. Some Special Signs in Chest Roentgenology. In C.B. Rabin (Ed.), *Roentgenology of the Chest.* Springfield, IL: C. C. Thomas, 1958.
8. Fleischner, F.G. The visible bronchial tree. A roentgen sign in pneumonic and other pulmonary consolidations. *Radiology* 50:184–189, 1948.
9. Felson, B. *Chest Roentgenology.* Philadelphia: Saunders, 1973.
10. Tew, J., Calenoff, L., and Berlin, B.S. Bacterial or non-bacterial pneumonia: Accuracy of radiographic diagnosis. *Radiology* 124:607–612, 1977.
11. Weill, H., George, R.B., Rasch, J.R., Mogabgab, W.J., and Ziskind, M.M. Roentgenologic appearance of viral and mycoplasma pneumonias. *Am. Rev. Respir. Dis.* 96:1144–1150, 1967.
12. Foy, H.M. Radiographic study of mycoplasma pneumoniae pneumonia. *Am. Rev. Respir. Dis.* 108:469–474, 1973.
13. Genereux, G.P., and Stilwell, G.A. The acute bacterial pneumonias. *Semin. Roentgenol.* 15:9–16, 1980.
14. Fraser, R.G., and Pare, J.A.P. *Diagnosis of Diseases of the Chest.* Philadelphia: Saunders, 1977.
15. Fletcher, B.D., and Avery, M.E. The effects of airway occlusion after oxygen breathing on the lungs of newborn infants. Radiologic demonstration in the experimental animal. *Radiology* 109:655–657, 1973.
16. Rahn, H. The role of N2 gas in various biological processes, with particular reference to the lung. *Harvey Lect.* 55:173–199, 1960.
17. Kattan, K.R., Eyler, W.R., and Felson, B. The juxtaphrenic peak in upper lobe collapse. *Semin. Roentgenol.* 15:187–193, 1980.
18. Golden, R. The effect of bronchostenosis upon the roentgen-ray shadows in carcinoma of the bronchus. *A.J.R.* 13:21–30, 1925.
19. Fleischner, F.G. Linear Shadows in the Lung Fields. In C.B. Rabin (Ed.), *Roentgenology of the Chest.* Springfield, IL: C. C. Thomas, 1958.

20. Hanke, R., and Kretzschmar, R. Round atelectasis. *Semin. Roentgenol.* 15:174–182, 1980.
21. Nordenstrom, B. New Trends and Techniques of Roentgen Diagnosis of Bronchial Carcinoma. In Simon, M., Potchan, E.J., and Le May, M. (Eds.), *Frontiers of Pulmonary Radiology.* New York: Grune, 1967, pp. 380–404.
22. Rigler, L.G. A new roentgen sign of malignancy in the solitary pulmonary nodule. *J.A.M.A.* 157:907, 1955.
23. Rigler, L.G., and Heitzman, E.R. Planigraphy in the differential diagnosis of the pulmonary nodule with particular reference to the notch sign of malignancy. *Radiology* 65:692–702, 1955.
24. Templeton, A.W., Jansen, C., Lehr, J.L., and Hufft, R. Solitary pulmonary lesions. Computer-aided differential diagnosis and evaluation of mathematical methods. *Radiology* 89:605–613, 1967.
25. Steele, J.D. *The Solitary Pulmonary Nodule.* Springfield, IL: C. C. Thomas, 1964.
26. Simon, G. *Principles of Chest X-Ray Diagnosis.* London: Butterworth, 1962.
27. Altemus, R. Localized interlobular septal lines in bronchogenic carcinoma. *J. Thorac. Cardiovasc. Surg.* 57:380–384, 1969.
28. Bryk, D. The participating tail. A new roentgenographic sign of pulmonary granuloma. *Am. Rev. Respir. Dis.* 100:406–408, 1969.
29. Hill, C.A. "Tail" signs associated with pulmonary lesions: Critical reappraisal. *A.J.R.* 139:311–316, 1982.
30. Rigler, L.G. A roentgen study of the evolution of carcinoma of the lung. *J. Thorac. Surg.* 34:283–297, 1957.
31. Nathan, M.H., Collins, V.P., and Adams, R.A. Differentiation of benign and malignant pulmonary nodules by growth rate. *Radiology* 79:221–232, 1962.
32. Nathan, M.H. Management of solitary pulmonary nodules: An organized approach based on growth rate statistics. *J.A.M.A.* 227:1141–1144, 1974.
33. Chinn, D.H., Gamsu, G., Webb, W.R., and Godwin, J.D. Calcified pulmonary nodules in chronic renal failure. *A.J.R.* 137:402–405, 1981.
34. Maile, C.W., Rodan, B.A., Godwin, J.D., Chen J.T.T., and Ravin, C.E. Calcification in pulmonary metastases. *Br. J. Radiol.* 55:108–113, 1982.
35. Dodd, G.D., and Boyle, J.J. Excavating pulmonary metastases. *A.J.R.* 85:277–293, 1961.
36. Hampton, A.O., and Castleman, B. Correlation of postmortem chest teleroentgenograms with autopsy findings. With special reference to pulmonary embolism and infarction. *A.J.R.* 43:305, 1940.
37. Castleman, B. Pathologic Observations on Pulmonary Infarction in Man. In Sasahara, A.A., and Stein, M. (Eds.), *Pulmonary Embolic Disease.* New York: Grune, 1965.
38. Fleischner, F.G. Observation on the Radiologic Changes in Pulmonary Embolism. In Sasahara, A.A., and Stein, M. (Eds.), *Pulmonary Embolic Disease.* New York: Grune, 1965.
39. Woesner, M.E., Sanders, I., and White, G.W. The melting sign in resolving transient pulmonary infarction. *A.J.R.* 111:782–790, 1971.
40. Westermark, N. On the roentgen diagnosis of lung embolism. *Acta Radiol.* 19:357–372, 1938.
41. Fraser, R.G., and Pare, J.A.P. *Diagnosis of Diseases of the Chest,* (2nd ed.). Philadelphia: Saunders, 1977.
42. Williams, J.R., and Wilcox, W.C. Pulmonary embolism. Roentgenographic and angiographic considerations. *A.J.R.* 89:333–342, 1963.
43. Keating, D.R. Thrombosis of pulmonary arteries. *Am. J. Surg.* 90:447–452, 1955.

44. Llamas, R., and Swenson, E.W. Diagnostic clues in pulmonary thromboembolism evaluated by angiographic and ventilation-blood flow studies. *Thorax* 20:327, 1965.
45. Fleischner, F.G., and Reiner, L. Linear x-ray shadows in acquired pulmonary hemosiderosis and congestion. *N. Engl. J. Med.* 250:900–905, 1954.
46. Trapnell, D.H. The differential diagnosis of linear shadows in chest radiographs. *Radiol. Clin. North Am.* 11:77–92, 1973.
47. Chang, C.H.J. The normal roentgenographic measurement of the right descending pulmonary artery in 1,085 cases. *A.J.R.* 87:929–935, 1962.
48. Heitzman, Jr., E.R., Fraser, F.G., Proto, A.V., et al. Radiologic Physiologic Correlations in Pulmonary Circulation. In *Chest Disease* (third series), Syllabus. Chicago: American College of Radiology, 1981, p. 375.
49. Keats, T.E., Lipscomb, G.E., and Betts, C.S. III. Mensuration of the arch of the azygos vein and its application to the study of cardiopulmonary disease. *Radiology* 90:990–994, 1968.
50. Preger, L., Hooper, T.I., Steinbach, H.L., and Hoffman, J.I.E. Width of azygos vein related to central venous pressure. *Radiology* 93:521–523, 1969.
51. Milne, E.N.C., Pistolesi, M., Miniati, M., and Giuntini, C. The vascular pedicle of the heart and the vena azygos. Part I: The normal subject. *Radiology* 152:1–8, 1984.
52. Pistolesi, M., Milne, E.N.C., Miniati, M., and Giuntini, C. The vascular pedicle of the heart and the vena azygos. Part II: Acquired heart disease. *Radiology* 152:9–17, 1984.
53. Milne, E.N.C., Imray, T.J., Pistolesi, M., Miniati, M., and Giuntini, C. The vascular pedicle and the vena azygos. Part III: In trauma—the "vanishing" azygos. *Radiology* 153:25–31, 1984.
54. Gammill, S.L., Krebs, C., Meyers, P., Nice, C.M., and Becker, H.C. Cardiac measurements in systole and diastole. *Radiology* 94:115–119, 1970.
55. Hoffman, R.B., and Rigler, L.G. Evaluation of left ventricular enlargement in the lateral projection of the chest. *Radiology* 85:93–100, 1965.
56. Wescott, J.L., and Ferguson, D. The right pulmonary artery-left atrial axis line. *Radiology* 118:265–274, 1976.
57. Higgins, C.B., Reinke, R.T., Jones, N.E., and Broderick, T. Left atrial dimension on the frontal thoracic radiograph: A method for assessing left atrial enlargement. *A.J.R.* 130:251–255, 1978.
58. Lane, Jr., E.J., and Carsky, E.W. Epicardial fat: Lateral plain film analysis in normals and in pericardial effusion. *Radiology* 91:1–5, 1968.
59. Carsky, E.W., Mauceri, R.A., and Azimi, F. The epicardial fat pad sign. Analysis of frontal and lateral chest radiographs in patients with pericardial effusion. *Radiology* 137:303–308, 1980.
60. Alavi, S.M., Keats, T.E., and O'Brien, W.M. The angle of tracheal bifurcation: Its normal mensuration. *A.J.R.* 108:546–549, 1970.
61. Haskin, P.H., and Goodman, L.R. Normal tracheal bifurcation angle: A reassessment. *A.J.R.* 139:879–882, 1982.
62. Baber, C.E., Hedlund, L.W., Oddson, T.A., and Putnam, C.E. Differentiating empyemas and peripheral pulmonary abscesses. The value of computed tomography. *Radiology* 135:755, 1980.
63. Kattan, K.R., Eyler, W.R., and Felson, B. The juxtaphrenic peak in upper lobe collapse. *Radiology* 134:763–765, 1980.
64. Rabin, C.B., and Baron, M.G. *Radiology of the Chest: Golden's Diagnostic Radiology Series,* section 3. Baltimore: Williams & Wilkins, 1980.

65. Rhea, J.T., DeLuca, S.A., and Greene, R.E. Determining the size of pneumothorax in the upright position. *Radiology* 144:733–736, 1982.
66. Kurlander, G.J., and Helmen, C.H. Subpulmonic pneumothorax. *A.J.R.* 96:1019, 1966.
67. Gordon, R. The deep sulcus sign. *Radiology* 136:25, 1980.
68. Heitzman, R.E. *The Mediastinum Radiologic Correlation with Anatomy and Pathology.* St. Louis: Mosby, 1977.
69. Genereux, G.P. The posterior pleural reflections. *A.J.R.* 141:141–149, 1983.
70. Goodman, L.R., Conrardy, P.A., Laing, F., et al. Radiographic evaluation of endotracheal tube position. *A.J.R.* 127:433–434, 1976.
71. Osmond, R.S., Rubenfine, M.D., Anbe, D.T., et al. Radiographic demonstration of myocardial penetration by permanent endocardial pacemakers. *Radiology* 98:35–37, 1971.
72. Hearne, S.S., and Maloney, J.D. Pacemaker system failure secondary to air entrapment within the pulse generator pocket: A complication of subclavian venipuncture for lead placement. *Chest* 82:651, 1982.

INDEX